CORONA
THE WORLD'S GREATEST FAKE OF THE TWENTY-FIRST CENTURY?

MAIK NOVY

Copyright © 2021 Maik Novy

All rights reserved.

ISBN:

INTRODUCTION

The "Corona crisis" shows what ignorance and false teachings can do through unreflective parroting on the part of virologists and epidemiologists. Of course, this "crisis" is a hit for businesspeople and scientists to position themselves in the limelight and continuously ignore the topic. Politicians and scientists who only refer to statistics have no idea about the whole matter. Still, television programs and podcasts, etc., pretend to the entire population that they have a picture of the whole thing. Statistics are not yet proof that something is as graphics represents it. Statistics do not provide any scientific statements about cause and effect, i.e., a causal relationship. Statistics can be manipulated at anytime and anywhere so that, for example, a pandemic never ends because the actual knowledge about viruses and bacteria is not available. And all the knowledge about viruses and bacteria that is already available is not knowledge if virologists and epidemiologists do not know what to do with it. For my part, in my free time, I only dealt with this topic and have concluded that everything that has been and has been publishing from a scientific point of view is bogus. Scientists who conscientiously do their work do not make such a fuss about a virus because my fellow human beings' dangers due to their stupidity are more significant for my life and others than any threat posed by viruses and bacteria. Not even June Almeida has ever published a dissertation on her Discovery - at least I could not find one, because she would have been obliged to do a dissertation on her Discovery at the time so that there are no contradictions in the statements about her Discovery - namely the coronavirus - would and could give. I recommend the videos:

"Wrestling for Truth" and "Profiteers of Fear," which only exist in German-speaking countries, so that you can get an idea of how industry influences science. Virologists and epidemiologists have no idea about genetics, geneticists who contradict themselves, confessed nature lovers on YouTube who have no idea about nature, and chemists who have no idea about chemistry. Still, all have an opinion and yet no idea.

For this reason, I have written down what I have learned about viruses and bacteria, in general, to be able to form an excellent overall picture for myself - regardless of the completeness and contradictions that naturally creep in and creep in when you have already been confronting with incomplete and contradicting statements. I have also found that the more I deal with this topic, the more contradictions and incompleteness become apparent to me. For this reason, I cannot and do not have to guarantee correctness and completeness. If you are interested enough readers, you will find out for yourself what is believable or not. Thank you for your interest, and I hope that I have provided enough clues for your self-critical view of the current pandemic. Should you come across questions that I have not yet answered, try to answer them with the already available and published readings. Remember that what you will read from me has not been officially or scientifically confirmed, and if so, only partially. Do not believe everything you hear or read - whether from me or others! Ask yourself why such a fuss is being made about a virus right now. Why not in the past? Why can virologists and epidemiologists not clarify this regard, e.g., the objective and the subjective difference between the SARS-CoV-1 and the SARS-CoV-2 virus? But the pandemic still has something positive. It shows how overwhelmed politicians, scientists, etc. are and how even the most remarkable rulers embarrass themselves.

Massive arsenals, nuclear weapons, etc., but fear of a virus - what a joke. A video on SARS from 2003, namely the Tagesschau from April 1, 2003, reported that the coronavirus contagion occurring in SARS originated mainly from Hong Kong, southern China, Hanoi, Singapore, etc. Of course, differences are now have made between SARS-CoV-1 and SARS-CoV-2. To date, no corresponding micrographic recordings have been publishing that show the objective difference between SARS-CoV-1 and SARS-CoV-2 because the circulating images are nothing more than 3D models that anyone can model and print at their discretion. For example, I mention the so-called torovirus, a coronavirus, but it has an elongated shape, changing it to become around coronavirus. Any virus can spread throughout the body via the bloodstream, settle somewhere, and, after a specific time, cause symptoms. One then has it with the intestines, the other with the lungs, and the head.

The coronavirus has been discovering because June Almeida was looking for the cause of rubella back in the 1960s and attributed it to the coronavirus. Unfortunately, I could not find a dissertation about it or from her. I have seldom heard so many stupid arguments and contradicting arguments from scientists as I did on the subject of "Corona." Claiming who is or has been vaccinating, there is still a risk of infection from that person who has not understood the principle of active protective vaccinations or the term. As soon as someone is vaccinated, the virus cannot spread at all. Otherwise, any type of immunization would be unnecessary because active or passive vaccination prevents the virus's multiplication and spread. There is only danger of a virus if it causes the related symptoms, which happens in about 24-72 hours, which I learned in biology class. So long as a corona-infected person has no signs, there is no danger at all. You can also use chicken eggs, for example, to find out whether an active ingredient affects living things or not. Take a whole hen's egg and inject the Vaccine into this hen's egg, which, by the way, is also done against salmonella. Incubate this egg in a breeder and give the chick injections of coronaviruses every seven days to determine which interval the chick starts to show symptoms. The breaks are synonymous with the cycle in a polymerase chain reaction. Please take a second chicken egg as a comparison, incubate it without a vaccine, and inject this chick with coronaviruses. As soon as this chick shows symptoms, give this chick the Vaccine and hope that it works; if it does not work the first time, give it at intervals until the Vaccine starts to work. This process can have been carrying out with all viruses and vaccines or individual RNA sequences. Instead of a vaccine, you give a hen's egg viruses - no matter which ones - and let them incubate in the breeder to determine whether viruses cause a genetic disposition. For example, you can give a vaccine and viruses simultaneously, but separately, to determine what happens then, for instance, whether vaccines have adverse effects on embryos. The whole thing can also have been carrying out with phages to determine whether phage therapy is an option to eliminate infectious diseases or genetic dispositions specified via the amniotic fluid.

Regarding the phages, it should also have been saying that, depending on the culture medium in which they are grown, they adopt the RNA sequence accordingly to carry out the corresponding reverse transcriptase to remedy genetic dispositions during pregnancy. For this, you need a Petri dish that, depending on the incorrect RNA sequence, has the nutrient medium, whereby the nutrient medium has the corrected RNA sequence. Phages have then been placing to stimulate reproduction under UV light and once under X-rays then to administer. Accordingly, a phage farm is set up, whereby two so-called breeders are required: once with UV light and once with X-rays. The breeders can also find out how viruses and bacteria behave. Two Petri dishes are always needed, both of which have the same culture medium, but one sample has constantly been exposing to UV light. One example has always been exposing to X-rays. Two neutral Petri dishes should always have been using as a reference. There is only water as a nutrient medium, but with identical viruses and bacteria to be examined, and here too, the UV light and X-rays should have been used once. Although bacteria are hydrophobic and behave like insects in water, water should still have used as a reference medium. There will not be any problems with viruses because they are hydrophilic anyway. Here now follows the law in memory of Carol Orzel (a woman with two skeletons) and the others who suffer or have suffered from the so-called FOP syndrome:

To do this, breed a bacteriophage or macrophage or both, whereby these must have a Golgi apparatus. It is then injected into one chicken egg to determine whether a second skeleton has formed and if so, we know why.

I want to add an RNA sequence so that it can have tested whether this sequence is responsible for the FOP syndrome:

$$
\begin{array}{lllll}
5'| & UU & GG & G|3' \\
3'| & G^{G} & GU & U|5'
\end{array}
$$

All in all, it does not matter whether I am right or not; the only thing that matters are that I am at least able to develop ideas: "If you only see the difficulties in the possibilities, you have to hide something sees the difficulty, he has something to say." The arrangement of the glycoproteins or amino acids of an RNA sequence or DNA sequence is random. For this reason, a primer with the same RNA sequence or DNA sequence cannot have developed at all since it is still not possible in genetics to isolate individual sections from an RNA or DNA sequence. Also, nature never creates an identical RNA or DNA sequence twice. Even with twins, there will be differences. Otherwise, all people would have to look the same in hair color, skin color, eye color, body size, facial features, and the like. It means that for everyone who has tested for COVID, all values should be negative or all positive. But with one positive and with others harmful, that should not be. It means that with every COVID test, every RNA or DNA sequence will always look different.

For this reason, Clemens G. Arvay is wrong when he says in his book: "Corona vaccines - rescue or risk?" (Only available in German) claims that specific glycoproteins or amino acids are always in the same place. I downloaded Mai Thi Nguyen-Kim's dissertation under her current surname Leiendecker from Wikipedia. As soon as I saw the number of pages, I knew immediately that experiments never bring as much new knowledge to light as what people have inclined to interpret into it, without any justification. Even Albert Einstein's dissertations were never this length but different. I claim that a dissertation that brings something new to light would be published accordingly in a specialist journal. But in the case of Mai Thi Leiendecker's dissertation, I am not aware of anything if I have not yet read it through thoroughly. But soon, I will study this dissertation.

At this point, it is worth mentioning the following law of hermeticism: "Whoever does not train his consciousness, the spirit working in him, and accordingly does not know how to use it, only unconsciously controls his circumstances and is not in a position to control them willingly influence." (From the book: "Hermetic" by Benedict Brefordt). A clumping of blood is not caused by the superimposition of individual blood platelets but by thrombocytes created when the cell lacks cytoplasm, causing blood to congest. White blood cells do not exist in this sense, but white blood cells are simply the blood plasma (yellow color) inside the cytoplasm cells and consist of cytosine. Whether the cytoplasm always has the yellow color is an open question at this point; it may well be that it can look different in other living beings, etc. The color component is used solely for the objective differentiation between the individual parts so that statements can be generalized more clearly and serve a higher degree of uniformity in conveying knowledge. Instead of the cytoplasm, uracil appears, which means that cells no longer divide. Uracil not only prevents cells from dividing but also prevents blood from clotting in wounds.

Nonetheless, I found a way to prove that a primer works when it obeys the following law:

Primer function

as the direction of winding up and

unwinding DNA and RNA sequences

to identify the origin of genomes of all kinds

Prof. Christian Drosten, Dr. Mai Thi Nguyen-Kim,

and Dr. Clemens G. Arvay

Equation:

$$f(\varpi) = \omega^{r^{\frac{1}{2}}}$$

Method:

1) Development of a DNA sequence!

2) After processing a DNA sequence, we get an RNA sequence!

3) This RNA sequence should then have rewound, whereby every cycle must match so that the DNA sequence in point 1) in every line is identical!

4) Application:

a) Development of a DNA sequence:

$$\begin{bmatrix}1 & \rightarrow & 2\\ \vdots & \ddots & \vdots\\ 3 & \rightarrow & 4\end{bmatrix}\begin{bmatrix}3 & \rightarrow & 4\\ \vdots & \ddots & \vdots\\ 1 & \rightarrow & 2\end{bmatrix}\begin{bmatrix}2 & \rightarrow & 1\\ \vdots & \ddots & \vdots\\ 4 & \rightarrow & 3\end{bmatrix}\begin{bmatrix}4 & \rightarrow & 3\\ \vdots & \ddots & \vdots\\ 2 & \rightarrow & 1\end{bmatrix}$$

b) After processing a DNA sequence, we get an RNA sequence:

1 2 34| 2 14 3| 34 1 2|4 3 2
1| ~

c) This RNA sequence should then have rewound, whereby every cycle must match so that the DNA sequence in point 1) in every line is identical:

$$\begin{bmatrix}4 & \rightarrow & 3\\ \uparrow & \ddots & \downarrow\\ 2 & \leftarrow & 1\end{bmatrix}\begin{bmatrix}2 & \leftarrow & 1\\ \downarrow & \ddots & \uparrow\\ 4 & \rightarrow & 3\end{bmatrix}\begin{bmatrix}3 & \leftarrow & 4\\ \downarrow & \ddots & \uparrow\\ 1 & \rightarrow & 2\end{bmatrix}\begin{bmatrix}1 & \rightarrow & 2\\ \uparrow & \ddots & \downarrow\\ 3 & \leftarrow & 4\end{bmatrix}$$

d)

1 2 43| 1 24 3| 12 4 3|1 2 4
3| ~

One should ask oneself why there are never any recordings of ambulances in motion or recordings of wards in hospitals in China when there are so many acute cases and deaths in China. Ask the Beijing residents how often they hear an ambulance or a police car driving with a siren. I can tell you, not once. Since the residents must ultimately pay their medical costs in China, their sickness rate is zero since no Chinese dares to see a doctor because that would be too expensive. Or ask a Chinese virologist whether he knows Robert Koch and his Koch postulates. Again,

I can tell you, not at all. Fatally, the Chinese only do technical or artistic courses in Europe, but hardly any medicine and the like, if not at all. Because of this, the hospitals are empty and have nothing to do with the wards.

The whole corona pandemic is a letdown. In everything that people are always to blame for, their illnesses can have ascribed to themselves through their way of life and behavior. With the words from Hermetics, from the book of the same name: "Hermetic" by Benedict Brefordt, it can be said: "People create their living conditions. Every person can change everything. Every person has their own life in their own hands. Become people sick; they are responsible for it. If people are poor, they are also responsible for it. Changing the way of thinking about the circumstances changes the way of thinking. Our consciousness, which is part of the total spirit, affects each of us and thus ensures a change in the world resulting from this total spirit." It also applies to the plague in the Middle Ages, which the people attributed to themselves through their wrong ways of life and behavior. This fact is not due to nature because illnesses are always a sign of unnatural life and conduct. Germany is one of the examples that lockdowns are useless. In the case of a "hard" lockdown, the incidence values go down but go up again after a lockdown. My fellow human beings also mistakenly suspect a deadly epidemic behind every disease since they can cope with their own meaningless, meaningless life.

Everyone wants to be something special these days, but they lack the necessary intelligence, skills, and abilities to create a new one out of their stagnant consciousness, as I do. I already know that without the whole corona pandemic caused by this "virus," there would never have been any severe difficulties worldwide. The full pandemic staged by politics and science serves only one purpose, to turn away from the significant issues. And all the vaccines are the only placebo anyway, and there will never be herd immunity in the first generation, but only with the next born. Even if everyone is vaccinated, every virus will adapt accordingly through phagocytosis so that the whole game could repeat itself forever. And even the game of lockdowns can have played on forever. Still, in the end, people are only sick because politics and science were never up to their tasks and were never up to their jobs since everything they do has only ever been misdirecting by lobbyists. For my part, I will exhaust this topic to not spoil my cheerful serenity - at least in my private life.

THE GENETIC STRUCTURE OF SARS-COV-2

Prof. Hendrik Streeck and June Almeida

Abstract:

The genetic structure of SARS-CoV-2 does rule an origin out of a laboratory. There is no severe acute respiratory syndrome, which has its source by a coronavirus. Genomic analysis can never show either what illness a patient has, or which virus has the appropriate genomic that could belong to this or that illness. Every geneticist would agree to that. Genetics has not been developed at all about the knowledge about viruses and bacteria. Accordingly, general knowledge about viruses and bacteria is neither available nor complete - and if so, only in individual cases. An analysis of bacteria and viruses makes sense if bacteria and viruses are isolated and then administered to a laboratory mouse. It is the only way to determine whether this bacterium or virus also belongs to the respective disease. The said acute respiratory syndrome also occurs in polio patients who are then dependent on an iron lung for the rest of their lives. Without viruses and bacteria, there would be neither vegetation nor life on this planet. It is not viruses or bacteria that cause disease, but deficiency symptoms manifest themselves, such as there is no clean drinking water or enough healthy food. Besides, since viruses have no metabolism, they cannot multiply either.

According to David Baltimore and Howard Themin, only retroviruses could use their RNA to reproduce and do not need a host for this. Everything else is nonsense since viruses with no metabolism of their own cannot replicate with a host's help. You can safely forget everything that has been written about viruses and bacteria so far.

Furthermore, I also claim that said coronavirus does not originate from bats. If a virus or a bacterium triggers a disease, it occurs first and foremost in animals such as foot and mouth disease, swine flu, bird flu, etc. Rossana Segreto and Yuri Deigin have no idea what they wrote in their essay. And what both wrote is just for the bin. Every Vaccine developed must first be tested for effectiveness, which has not been done with any vaccines developed to date. It is, of course, irresponsible on the part of politicians, doctors, and virologists. Before testing a vaccine, a virus must first have injected into a laboratory mouse. The disease must break out in the laboratory mouse. The Vaccine must then have administered to determine how much of a vaccine is needed and how often vaccines can have given to eradicate the disease. Also, this process must have been carrying out for each Vaccine. An analysis of a virus is not possible with the PCR test because it is much too tiny.

For this reason, the required amount of material is not available to use a PCR test to analyze the RNA of a virus for being able to reproduce would happen to generate a DNA. Even as a layman, I have now understood what has been reporting in the documentation about DNA traces in forensics, that although only tiny amounts of DNA material are required to be able to carry out a PCR test, the amount of material that a virus brings with it would be unsuitable even for the PCR test, since the size of a virus is in the millionths of a millimeter - so to speak, in an area of light that is no longer even visible. Besides, if a virus were to be grown in a laboratory, an employee would have to become infected with a virus to carry it out with him to spread. Acute respiratory diseases only occur connected with coronary artery disease or pneumonia, etc., if the vessels have already calcified at an advanced stage. However, this has been excluding from the outset, which suggests incompetence on doctors and virologists.

Furthermore, PCR tests have only been using to replicate RNA or DNA but not analyze viruses or bacteria. Since a reference must be available beforehand to compare RNA and DNA, it can have been using as a guide to avoiding misinterpretations. The corona pandemic shows politicians, doctors, and virologists' natural faces, namely that they have no sense of responsibility and only spread lies. There is no difference between ribonucleic acid and deoxynucleic acid. The only difference between RNA and DNA is that the RNA is a single strand, i.e., sRNA, and the DNA is a double strand, i.e., ssRNA. The polarity has only been using to distinguish whether it is of male or female origin. One speaks of negative polarity if the basis is female; if the root is male, one speaks of positive polarity. The contradiction has also been using to determine which nucleic acid has converted to read from left to right with a positive polarity and read from right to left with a negative polarity. For example, ABCD is the positive polarity, and DCBA is the mirror image. And in this case, one speaks of an mRNA, in which the standard ensures that A has converted into D, B into C, C into B, and D into A to compensate for a deficiency, if it is present. What humans call the immune system is nothing more than metabolism, and there are no T-helper cells. The metabolism is subject to the natural law of thermodynamics. This law requires that an organism absorbs substances and converts them since the natural law of thermodynamics by Henri Poincare is about a material interaction or absorbing an essence and then converting it to provide the required amount of energy generated.

In physics, thermodynamics' natural law has been using to create general field equations that conserve energy-mass-momentum. The SARS-CoV virus, if it is a virus at all, belongs to the class of the so-called Orthomyxoviridae under the premise that it is a virus. The family of the Orthomyxoviridae (Greek: "myxa," German: "Schleim") includes enveloped viruses with double-stranded RNA, as they have a double lipid membrane shell, with positive and negative polarity. That means with male and female parts. The Orthomyxoviridae include virus genera that infect and multiply in the respiratory system mainly via droplet infection. Besides, the Thogotovirus triggers the so-called Ebola. The thogotovirus can also cause AIDS. This virus becomes a host in a tick and reproduces, and a tick turns into a beetle. In general, a mutation occurs when there is no longer any cell division, which means that one must first prove that viruses prevent cell division, so to speak, withdraw the organism's proteins to obtain the necessary material for transformation. It should also have added that every virus must produce a parasite. How contagious a virus is must first have been determining under laboratory conditions and how fast the virus replicates in a Petri dish. Because the faster a virus multiplies in a Petri dish, the greater the risk of infection. Without cultural viruses in a Petri dish, it is impossible to determine how dangerous a virus is. Every vaccination will ultimately lead to a virus adapting accordingly. The virus does not care where it gets its material because, preferably, each virus uses only its RNA to replicate itself. If a virus were to use a different RNA, there could no longer be any replication question but the only mutation. For a virus to continually reproduce and multiply, it may just use its RNA. The difference between replication and mutation lies in those viruses can only be replicated if they only use their RNA for reproduction. Still, the transformation is referred to when they use foreign RNA. Without viruses, no living being could adapt to the changing living conditions. Coronaviruses get the energy they need from nitrogen, which increases the uracil content in the cells so that cell division in the host has been suppressing. It creates the risk of inflammation, which then leads to cancer. Physical activity reduces the nitrogen content, producing more heat, and puts the coronavirus into a state of phagocytosis. Virus infections arise when they penetrate the cell to reproduce and remain there until it no longer provides enough space and nutrients for reproduction, which forces them to leave the cell in question to colonize other cells. A mutation of cells that develop into cancer cells has also been causing by increased uptake of pollutants or an increased lack of nutrients, such as tobacco smoke, industrial and car exhaust fumes, pollution in the workplace, or a lack of fresh air own four walls. But also, through alcohol, radioactive substances, illegal drugs, in short, everything that causes the body to have been forcing to destroy itself.

To determine the size of a virus, you need a ruler with 0 to 100 nanometers, each 1 nanometer. It means that every single line of the ruler should be a millionth of a millimeter. Still, the bar itself may be less than a millimeter, preferably a tenth of a millionth of a millimeter, but this is not possible because the nanometer range is invisible to us humans. Therefore, we humans cannot manufacture such a measuring tool as a ruler with such a division. But that also means that we cannot determine the size of a virus. I sincerely ask you to remove a 60 - 120 nanometers crystal from a teaspoon of sugar with a pink chain help. You will not succeed. Because of this, I know, so it is a statement that the PCR method of determining the RNA of a virus fails, must fail, and most importantly, will fail. Any experienced geneticist will tell me that.

For this reason, I also claim that the whole pandemic is just due to the Bullshit of virologists. Anyone who wants to accuse me of incompetence, precisely because I am not a scholar, but only a person with a negative perception, will not make it. Without knowing about the Special and General Theory of Relativity, which I received from Albert Einstein and others through their works, I would not determine all of this.

Plants, animals, and humans also have the so-called prickly protein. Without the sting protein of a coronavirus, many plants and animals would not have a sting. However, it looks different with humans because they would have no hair without the coronavirus's spiky protein. Also, the spike protein of the coronavirus has only been producing by guanine. Viruses, also known as macrophages, cannot produce spines, but bacteria, also called bacteriophages.

In evolutionary terms, bacteriophages cause the Y chromosome, and macrophages induce the X chromosome's presence.

I want to define the difference between bacteriophage and macrophage below:

The bacteriophage and thus the Y chromosome consists of guanine, thymine, cytosine, and adenine.

The macrophage and thus the X chromosome are guanine, cytosine, uracil, and adenine.

The guanine is not the cause for the formation of so-called peplomers, but the real reason lies in the Coriolis forces' effect.

However, guanine is the cause for the formation of so-called telomers (pili).

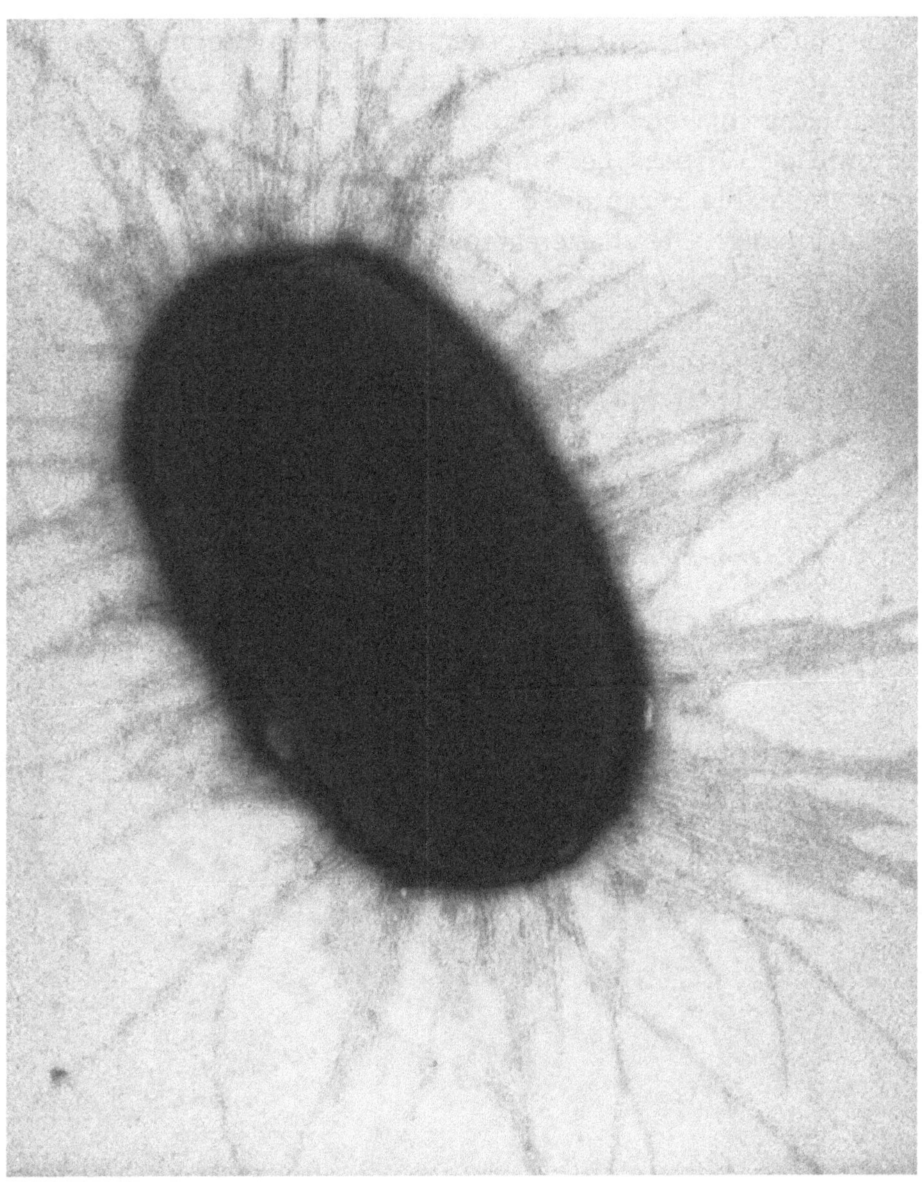

Quelle: Datei: E. coli fimbriae.png - https://de.wikipedia.org

Besides, the macrophage (large phagocytes) eats the bacteriophage (small phagocytes). As a result, it creates jelly-like sputum, which is also known as mucus. This fact is also known as phagocytosis.

Microorganisms with telomeres on the surface are called bacteria (bacteriophages). Microorganisms with peplomers on the body are called viruses (macrophages). Viruses are unicellular organisms that enclose bacteria because bacteria need a host to multiply. Viruses, i.e., single-celled organisms, do not need a host to be able to reproduce. When a macrophage surrounds a bacteriophage, DNA has created. Viruses are always hydrophilic, whereas bacteria are aerobic. Because of this, the bacteria need a host cell. The likelihood that I can trash what I have written so far increases from time to time. It means that it is not the viruses responsible for many diseases, but only the bacteria.

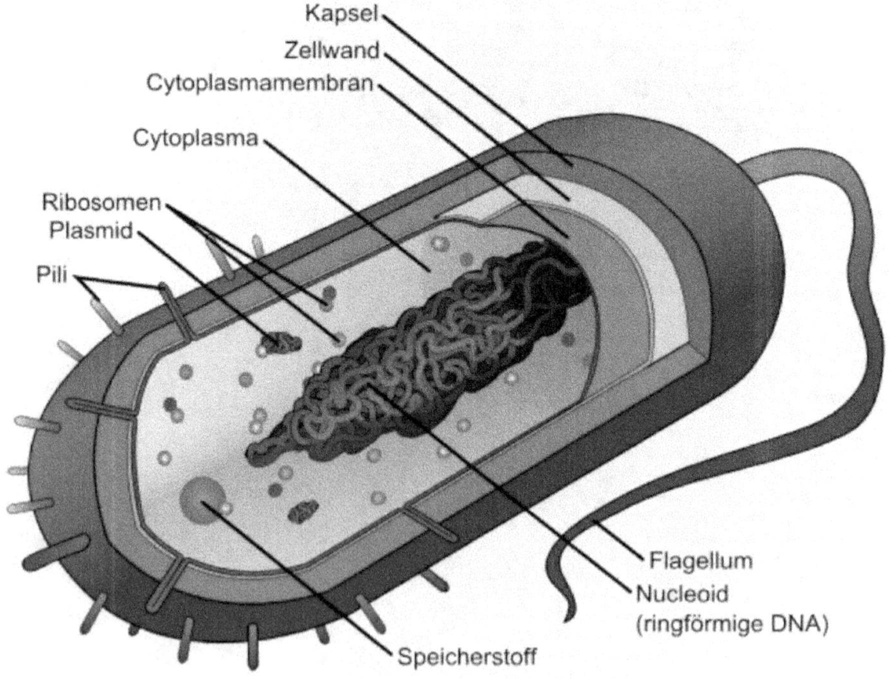

Quelle: Datei: Average prokaryote cell - de.svg - https://de.wikipedia.org

Quelle: Datei: Animal cell structure de.svg - https://de.wikipedia.org

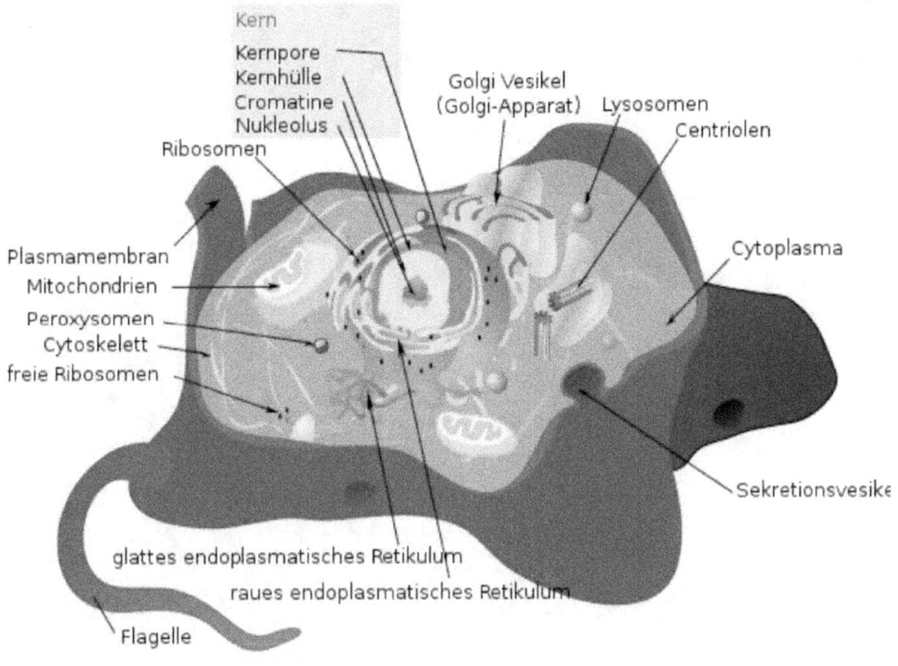

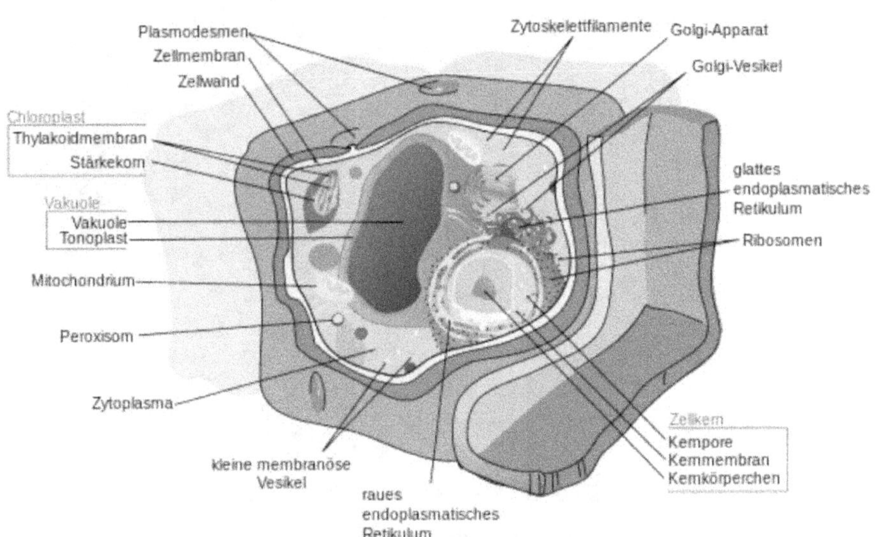

Quelle: Datei: Plant cell structure svg-de.svg - https://de.wikipedia.org

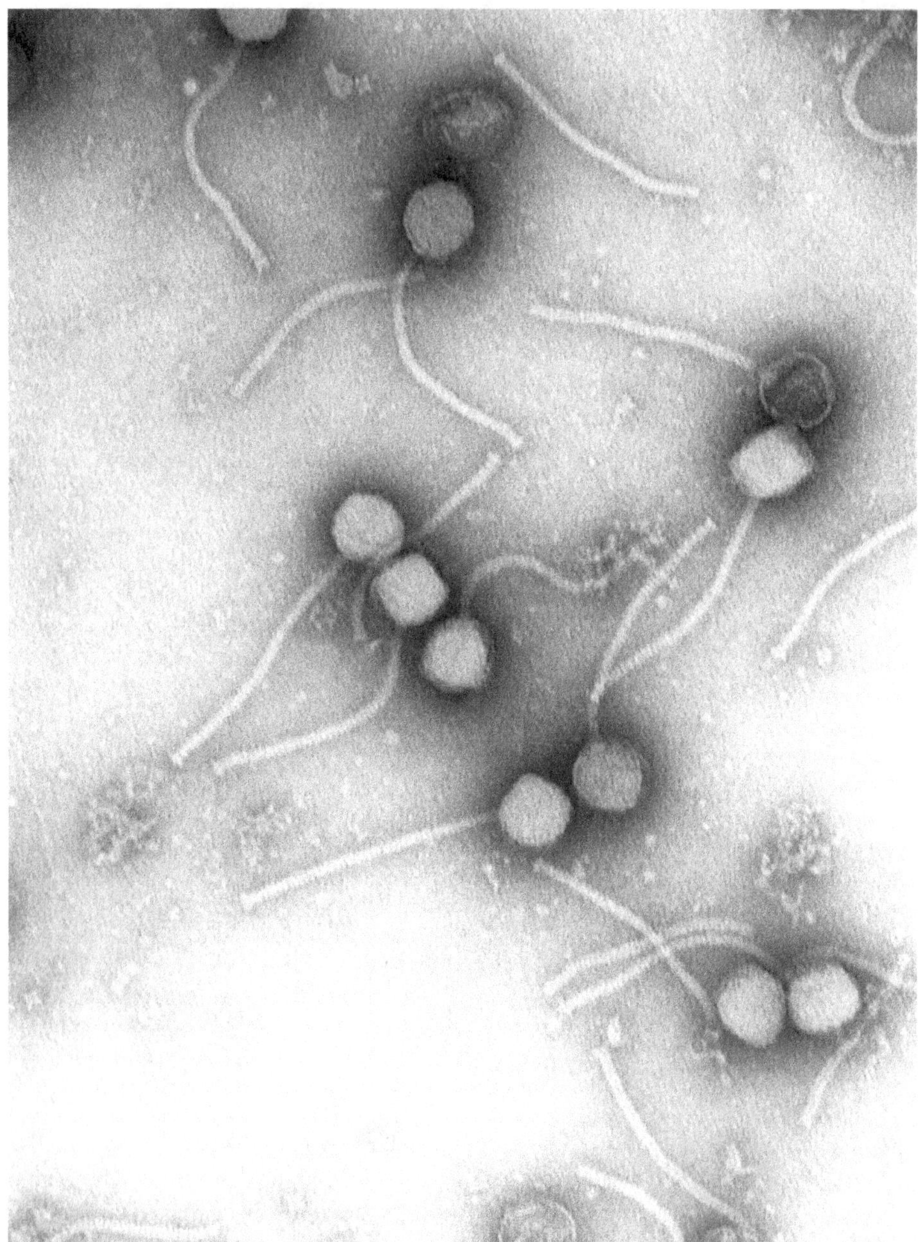

Quelle: Datei: Gamma phage.png - https://de.wikipedia.org

Bacteriophages are bacteria that need viruses as host cells.

The jellyfish is the best example that a living being can grow and multiply without metabolism or a reproductive organ. In the jellyfish, one can see very well what telomeres do. With the help of telomeres, bacteria penetrate the viruses and force them to divide. It means that the sperm thread ensures that the ribosome (sperm) can penetrate the egg cell and stimulates the division through the sperm thread.

Invisibility begins in the nanometer range because as the electrons' speed increases, the light intensity decreases.

The AstraZeneca vaccine does not cause inflammation in the spinal cord's bone marrow. Patients with multiple sclerosis only have numerous sclerosis cases because their spinal cord bone marrow is inflamed, leading to paralysis, as was the case with Stephen W. Hawking. For Curing multiple sclerosis, the AstraZeneca vaccine must be injected directly into the spinal cord's bone marrow. Adenoviruses create the same effect as the body's retroviruses in chimpanzees and humans. I came up with it because I started reading the book: "Corona Vaccines - Rescue or Risk" by Clemens G. Arvay. Viruses with peplomers on the surface are basically what we call antibodies. These so-called enveloped viruses are all DNA viruses that only need uracil to divide. However, one should be careful, as viruses can also cause paralysis through their peplomers. Viruses with peplomers on the surface are poxviruses, adenoviruses, polioviruses, etc. Viruses with telomeres on the body are MERS viruses, etc., where MERS viruses cause measles, leaving people with facial scars once the measles clears up. Viruses with telomeres on the surface are RNA viruses that bacteria ingest to divide by themselves. Viruses are always species-specific. Depending on the virus, the emergence of the corresponding species. The contradiction in the whole dilemma is that it has claimed that viruses cannot reproduce without a host, but it is not. If that were the case, there would be no basis for the existence of plants, animals, and humans.

Furthermore, the difference between viruses with telomeres on the surface and peplomers is that enveloped viruses require UV light, whereas non-enveloped viruses require X-rays. I would suggest the following approach when it comes to analyzing RNA and DNA. I want to take the harvester as an example. First, cut off the legs and grind them in a mortar. The body must also have ground in a battery. To have a reference at all, one takes a vessel with only water, then a ship where the crushed legs are in it without any additive; this should also have done with the crushed body. Then all three samples are analyzed one after the other.

The harvestman legs then supply the RNA, the body, the DNA, and the water as a medium for dissolving the necessary reference. The RNA and DNA sequence can have been correctly interpreting when using water as a solution. However, RNA and DNA analysis can lead to falsified results if a foreign mRNA operates as the medium. If you want to use mRNA instead of water, you must know the RNA first. However, this is only possible if it has been analyzing without any additions. If it is available as a reference, the RNA can then be used as mRNA to exact an exact copy of the referenced RNA. It means that a DNA sequence is then present in which the first RNA single strand has read from left to right and the copy of this RNA single strand - mirrored - from right to left. To find out how non-enveloped and enveloped viruses arise, a chemist, for example, can put the given chemical elements of RNA and DNA in a Petri dish and irradiate them once with X-rays and once with UV light dispensing with the element water shall be. Also, you can do the same with water or hydrogen to see the reactions. If the chemical elements of RNA and DNA are exact, then results should come about. If this is not the case, it must have been clarifying which conditions must exist. The whole thing can also have been carrying out with mRNA active ingredients by taking any mRNA active ingredient and irradiating it once with and without water or hydrogen with UV light and X-rays to determine whether mRNA is an active ingredient is causing it whether viruses emerge from it.

To develop new species, one only needs to use the PCR method to analyze the chemical elements. It means that each chemical element has given an RNA sequence, i.e., transcribed, to then develop a corresponding DNA with the RNA sequence's help. Thus, new species of animals, plants, and humans emerge from it.

Segreto-Deigin method for the determination and control of RNA and DNA analyzes:

<u>Bacteriophage (Y chromosome)</u>

(- as an example -)

1)

$$RmYN02 = period = T = \frac{2\pi}{\omega}$$

In the determination of the period of RNA and DNA, all four lines have lined up.

Condition:

$$\begin{bmatrix} G & \rightarrow & A \\ \vdots & \ddots & \vdots \\ A & \rightarrow & G \end{bmatrix} \begin{bmatrix} C & \rightarrow & T \\ \vdots & \ddots & \vdots \\ T & \rightarrow & C \end{bmatrix}$$

$$\begin{bmatrix} C & \rightarrow & T \\ \vdots & \ddots & \vdots \\ T & \rightarrow & C \end{bmatrix} \begin{bmatrix} G & \rightarrow & A \\ \vdots & \ddots & \vdots \\ A & \rightarrow & G \end{bmatrix}$$

Annotation:

The direction of the arrow indicates the respective reading direction, disregarding the dashed lines.

Result:

$$\begin{array}{cccccccccc} G & A & CT| & A & GT & C| & CT & G & A|T & C \\ & & A & G & & & & & & \end{array}$$

2)

$$Query = Angular\ frequency = \omega = \frac{2\pi}{T}$$

In determining the RNA and DNA, the corresponding sequence must now determine using the angular frequency.

Condition:

$$\begin{bmatrix} G & \rightarrow & A \\ \uparrow & \ddots & \vdots \\ A & \cdots & G \end{bmatrix} \begin{bmatrix} C & \rightarrow & T \\ \vdots & \ddots & \downarrow \\ T & \cdots & C \end{bmatrix}$$

$$\begin{bmatrix} C & \cdots & T \\ \uparrow & \ddots & \vdots \\ T & \leftarrow & C \end{bmatrix} \begin{bmatrix} G & \cdots & A \\ \vdots & \ddots & \downarrow \\ A & \leftarrow & G \end{bmatrix}$$

Annotation:

The arrow's direction also determines the reading guide, and the dashed lines are negligible.

Result:

$$G \quad A \quad C|T| \quad \underset{C}{C} \quad \underset{|T|}{A|G|} \quad A \quad C|T| \quad C \quad A|G| \quad A$$

The respective letter between the lines should signal the transition and appear only once, but not twice in a row.

3)

$$Subject = Phase\ shift\ =\ diagonal = \varphi$$

This time the sequence must result from the diagonal to determine the RNA or DNA. The arrow's direction also determines the reading approach, and the dashed lines are not considered.

Condition:

$$\begin{bmatrix} G & \cdots & A \\ \vdots & \searrow & \vdots \\ A & \cdots & G \end{bmatrix} \begin{bmatrix} C & \cdots & T \\ \vdots & \searrow & \vdots \\ T & \cdots & C \end{bmatrix}$$

$$\begin{bmatrix} C & \cdots & T \\ \vdots & \searrow & \vdots \\ T & \cdots & C \end{bmatrix} \begin{bmatrix} G & \cdots & A \\ \vdots & \searrow & \vdots \\ A & \cdots & G \end{bmatrix}$$

Result:

$$T| \quad C \quad \begin{matrix} C|A \\ C| \end{matrix} \quad \begin{matrix} T \\ T \end{matrix} \quad A|G \quad G \quad GG| \quad A \quad TA| \quad C$$

4)

$$ZC45 = frequency = \frac{1}{T} = \frac{\omega}{2\pi}$$

This time it is a little more complicated because now you must shuffle the rows with each other. For this reason, I will have to design the division differently so that the whole thing is and remains comprehensible.

Condition:

The first and third lines are linked together.

$$\begin{bmatrix} G & \rightarrow & A \\ \vdots & \ddots & \vdots \\ & \cdots & \end{bmatrix} \begin{bmatrix} C & \rightarrow & T \\ \vdots & \ddots & \vdots \\ & \cdots & \end{bmatrix}$$

$$\begin{bmatrix} C & \rightarrow & T \\ \vdots & \ddots & \vdots \\ & \cdots & \end{bmatrix} \begin{bmatrix} G & \rightarrow & A \\ \vdots & \ddots & \vdots \\ & \cdots & \end{bmatrix}$$

The second and fourth lines are linked together.

$$\begin{bmatrix} & \cdots & \\ \vdots & \ddots & \vdots \\ A & \rightarrow & G \end{bmatrix}\begin{bmatrix} & \cdots & \\ \vdots & \ddots & \vdots \\ T & \rightarrow & C \end{bmatrix}$$

$$\begin{bmatrix} & \cdots & \\ \vdots & \ddots & \vdots \\ T & \rightarrow & C \end{bmatrix}\begin{bmatrix} & \cdots & \\ \vdots & \ddots & \vdots \\ A & \rightarrow & G \end{bmatrix}$$

Result:

$$G \quad A \quad CT| \quad A \quad GT \quad C| \quad CT \quad G \quad A|T \quad C$$
$$A \quad G$$

<u>Calculating the Taylor polynomial:</u>

$$Period\ multiplied\ by\ frequency = time\ interval$$

$$as\ a\ Taylor\ polynomial$$

$$Taylor\ polynomial:$$

$$T(\varphi,\omega) = \frac{2\pi}{\omega}\, x \frac{\omega}{2\pi} = 1$$

function of a time intervall:

$$f(x(t)) = \frac{2\pi}{T}$$

Cataracts in newborns have not been causing by rubella but by alcohol during pregnancy. A cataract is better known as a cataract (German: "Grauer Star"). Rubella embryo fetopathy occurs only when kidney failure during pregnancy leads to shingles in the expectant mother. I am becoming more and more confident that my way of life and behavior lead to deficiency symptoms responsible for diseases. I also notice that there are no electron microscopes or transmission electron microscopes in official laboratories. If viruses are in the nanometer range, how come simple microscopes have still used?

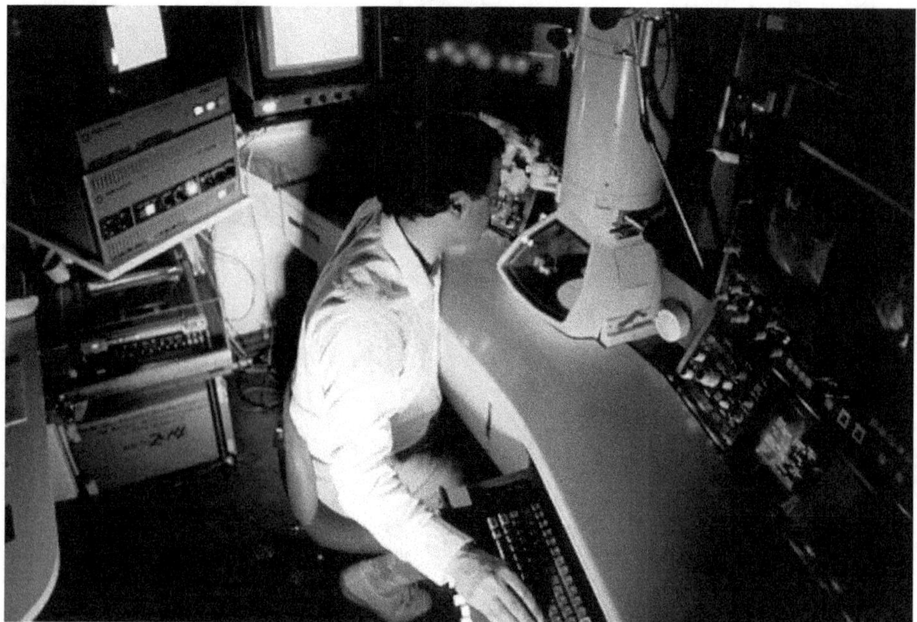

Quelle: Datei: CSIRO Science Image 410 Transmission Electron Microscope.jpg - https://de.wikipedia.org

There are no autoimmune diseases. The term has only been using as an excuse for personal misconduct. For example, destroying oneself through drugs and alcohol and the like and then looking for reasons for one's physical, mental, and spiritual ailments is poor. Or refuse to eat and then wonder if the body decomposes, i.e., destroys itself, because a lack of nutrients demands this or overeating and then complaining about cardiovascular problems. It is only the tip of the mountain idiocy society. Virologists behave just as stupidly and do not even know that you must inject a vaccine directly into a target tissue because hoping that you inject a vaccine into the shoulder that will eventually get into the lung tissue only points to incompetence - because of impotence and frigidity - the scientists. I can jerk off human RNA in semen or vaginal secretion for swallowing and have created a symbiosis.

According to Dr. Mike Hansen, the law says the following: 1) Every living being, and plant already has coronaviruses in them. In this case, one speaks of latency. Latency occurs when a virus is present but does not cause disease by itself. A disease caused by a virus already present in latent Form only occurs if the said virus serves as a host for a bacterium. For this reason, one speaks of a bacteriophage when a bacterium uses a virus as a host and may even need it. It is also why coronaviruses can also have been detecting in people who have no symptoms. As already mentioned, a virus does not cause any disease by itself and certainly not by itself, since without viruses and bacteria, there would be no life. 2) A virus can produce carcinogenic properties in cells when a bacterium inside a virus starts to grow that virus in a host cell, causing the host cell to divide into an adverse effect. Carcinogenic properties can also occur when reverse transcriptase is no longer possible. It means when a material interaction no longer takes place, viruses can implant themselves in dead host cells to make then them grow. It can also happen if antibodies attack a virus if it already contains a bacterium. 3) Lytic properties of a virus are given when a bacterium occupies a virus, whereby it triggers diseases, provided that the bacterium has so-called pili, whereby the virus then becomes a so-called non-enveloped virus. 4) Phagocytosis is supposed to protect the body from this, not only from bacteria but also from other pollutants, etc. Phagocytosis serves to subject the body to reverse transcriptase, so that cell division remains active if one stays active and provides sufficient rest and food, and the like. Phagocytosis is when a virus ingests a bacterium so that the virus can divide at all, so it forms mucus to renew dead cells at all.

Professor Christian Drosten rightly claims that viruses exist in the latent Form, not to produce any symptoms. The reason is that, e.g., coronaviruses with peplomers on the surface are why humans and animals have, e.g., toes and fingers, and men have a penis, etc. But Professor Hendrik Streeck also rightly claims that we must learn to deal with every type of virus and bacterium. And we should stop spreading such scare tactics. Ultimately, the whole scare tactic about coronaviruses is entirely superfluous. And to display such a ruinous attitude only shows the ignorance of scientists, or instead, it is probably due to the lack of interest in said scientists to do their work. Of course, every scientist has the disadvantage that he depends on research fellows, so that every scientist runs the risk of selling his ideas to the most bidder. Moreover, scientists are freelancers, so they usually must stay afloat with the said scholarship holders or by selling their non-fiction books, unless they have been giving a professorship or assistant position at a university or a permanent contract with affiliated companies to research their interests and spread it, because companies only support research to benefit themselves, no matter what the cost.

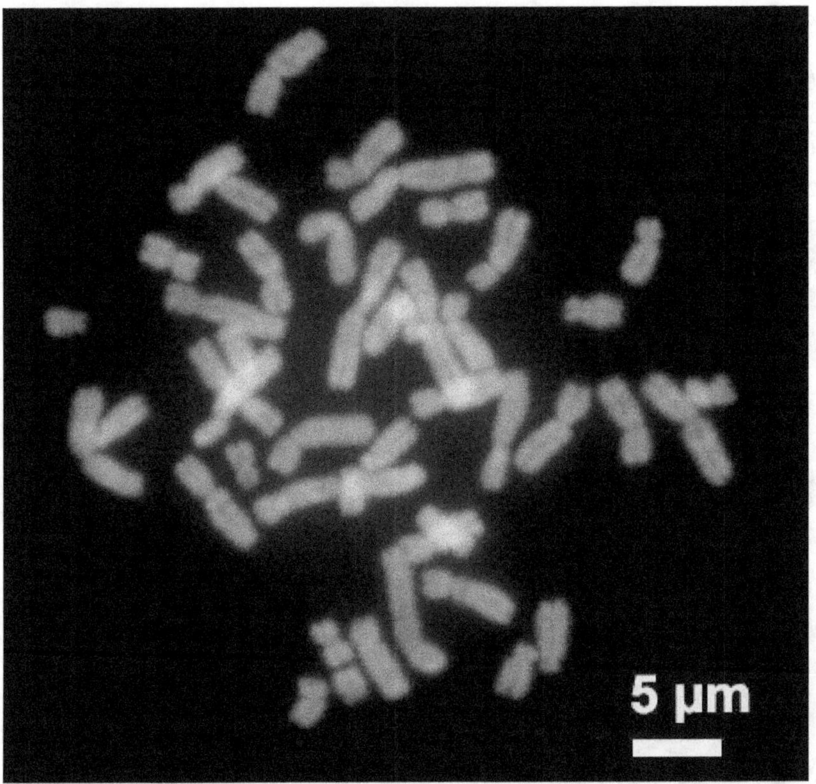

Quelle: Datei: HumanChromosomesChromomycinA3.jpg - https://de.wikipedia.org

Someone should explain how to recognize the genetic structure and elements in a blurred image like the one above. So much for if more have interpreted than can be observed. I cannot see X chromosomes, Y chromosomes, or any double helixes. Also, in the corresponding source's directory and the associated page, I did not find the famous "Photo 51", which Rosalind Franklin has ascribed. There are no photos, only graphical representations drawn because you cannot enlarge them enough to picture. Of course, you need a beginning to get ahead in research, but to claim to have observed things that cannot have been observed at all, since these can even have been recording with an electron microscope, I find that very bold. Suppose double helixes exist, and you can penetrate the nanometer range with the help of transmission microscopy. Why has an original image of a double helix never been produced, and even Wikipedia cannot help in this case. I am, of course, happy to be taught a better thing in this regard and at any time. In the picture above, you can see highlighted areas in the chromosomes. These are namely the Y chromosomes. Y chromosomes contain telomeres and are therefore Y chromosomes.

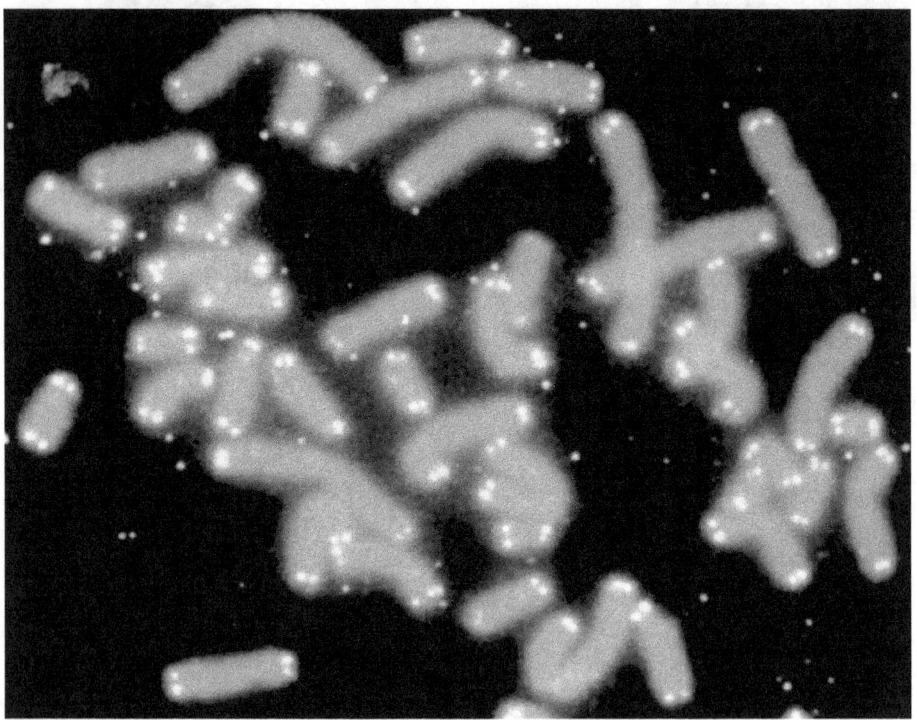

Quelle: Datei: Telomere caps.gif - https://de.wikipedia.org

The X chromosomes have been marking with 5' and the Y chromosomes with 3', but this has not been explaining in the context.

Plants, like viruses, are not living beings in the strict sense of the word, but only "close to life," yet these substances must have been absorbing to thrive. On the negative side, I must object that Prof. Christian Drosten and Prof. Hendrik Streeck spread their ignorance. You can usually tell from the formulations chosen, and in every program, be it as a podcast or TV show, they all always talk about the same superficial booze, that is, around the bush. In this way, however, they only achieve that mistrust grows in the population since what they say is neither explained in more detail nor specified and does not clarify. Because of this, not only I but also other outsiders tend to contradict a current topic such as the corona pandemic.

Worrying about side effects is completely exaggerated. Since the mRNA used is based on a purely biological basis anyway, side effects have been excluding. A virus is harmless to all biological systems and is much more necessary than a strictly natural mRNA. Since all virologists and epidemiologists, and biologists have no idea about evolution, they will always doubt the gifts of nature and see nature not as a friend but as an enemy, because you cannot make money with healthy people, so you talk people about a disease one and people are so stupid and fall for it, maybe because they do not feel like working. People whine about the potential side effects of vaccines, but when it comes to getting drunk or engaging in other abuse types, they do not care about the consequences. Enveloped viruses supply the body with nutrients so that an ATP synthase (adenosine dehydrogenase) is possible in the first place. In the warmer months and the longer days, the number of enveloped viruses is more significant than in the colder months. Since the number of enveloped viruses decreases in the colder seasons, this is also why plants and sometimes animals go into hibernation. You can recognize an ATP synthase in plants by the fact that they start to bloom. Animals begin to reproduce, and humans then start to complain.

There are no white blood cells. The blood's so-called plasma is nothing more than cytosine (yellow), located inside a compartment. The cell membrane is made up of thymine and is the reason the blood is red. The whole corona pandemic is fake because the Chinese scientist (name-dropped!) only started a rumor, and the entire world fell for it. Since the corresponding publication by the Chinese shows a lot of contradictions and incompleteness. I do not even need to read that; I know that. The mRNA is nothing more than the DNA that has been introducing into the body. In women, the egg cell provides the DNA.

On the other hand, men have two components: the RNA of the sperm thread and the DNA of the sperm cell or the sperm cell, also known as the ribosome. The sperm cells (ribosomes) trigger reverse transcriptase inside the egg by connecting them with the help of their sperm thread. So, the egg cell's mRNA (DNA) consists of GATC. And the sperm cell mRNA from GATC and that of the sperm filament from GACU.

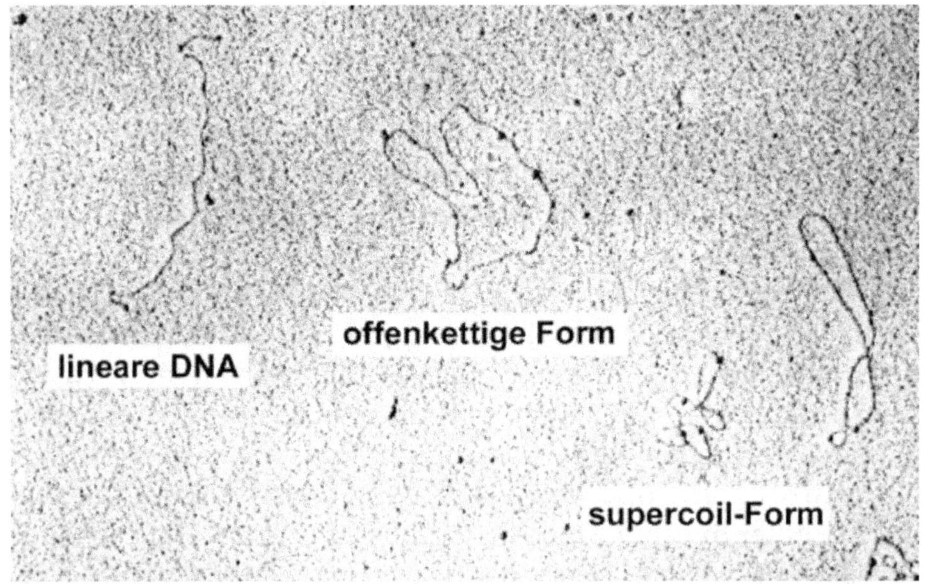

Quelle: Datei: Plasmidem.jpg - https://de.wikipedia.org

The picture above makes it difficult to distinguish between DNA and RNA, proving what interpretations are suitable for when the corresponding images are blurred.

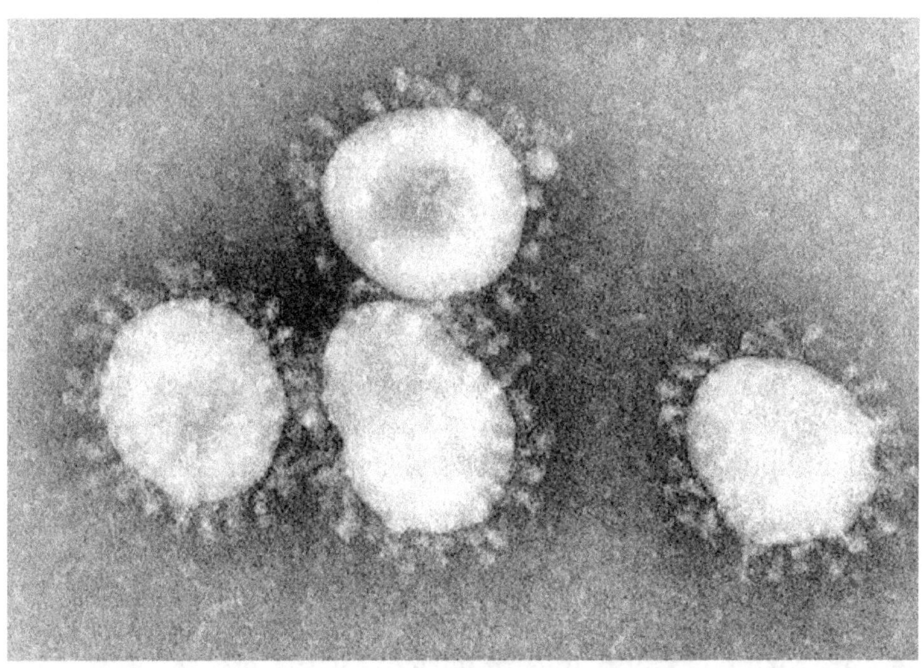

Quelle: Datei: Coronavirus 004 lores.jpg - https://de.wikipedia.org

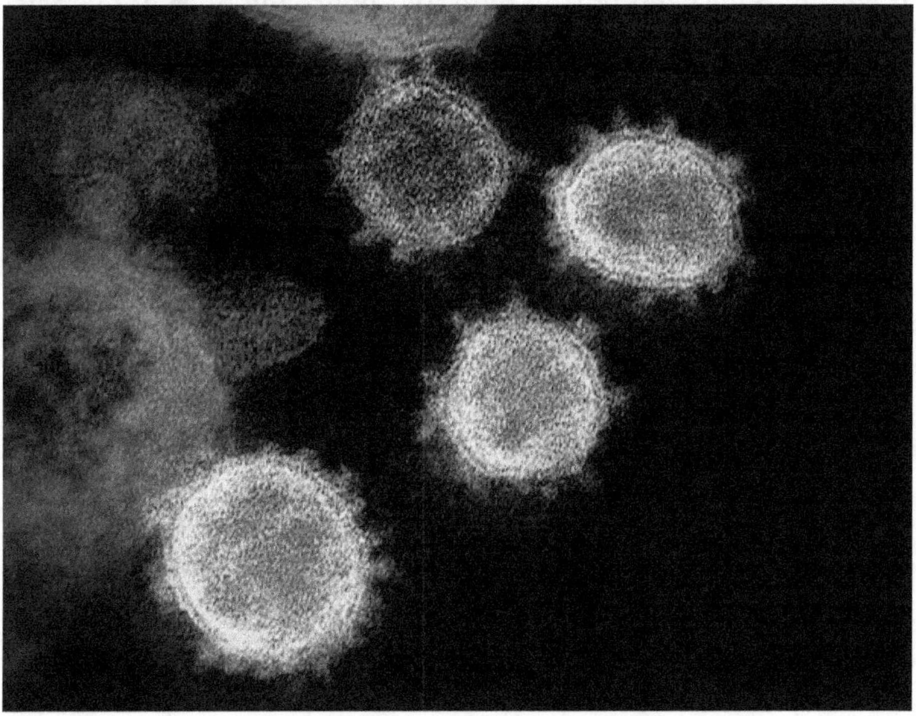

Quelle: Datei: Novell Coronavirus SARS-CoV-2.jpg - https://de.wikipedia.org

In the above images, more can see very well that a pure electron microscopic image can only recognize purely objective differences if mutations occur. But, on the other hand, you can even read the mRNA in the TEM image, consisting of thymine, cytosine, adenine, and guanine - at least if you use the picture as a template.

Coronaviruses can absorb substances to multiply by removing the cytoplasm from the cells so that circulatory disorders arise. In the series: "Stargate Atlantis," you can experience very well what happens when the Wraiths use coronaviruses to deprive people of their cytoplasm or what happens when the human organism loses more and more of its cytoplasm. Without cytoplasm in the cells, the human body begins to age. Besides, as soon as the cytoplasm disappears from the cell, so-called platelets are formed responsible for varicose veins, etc.

It should have been noting here that hydrogen does not consist of helium and oxygen but oxygen and nitrogen. The evidence has been providing by the fact that there are coral reefs and crabs. Hydrogen, like water, cannot consist of a mixture of helium and oxygen because helium is naturally a flammable liquid; according to Antoine Lavoisier, oxygen is combustible air, which should also have been mentioning here. And when two volatile elements have mixed, you get another flammable feature. The helium-oxygen mixture was used in human rocket engines, but the high reaction time means that this mixture was quickly exploited. And that is why I am right, as I have found that the water we drink, also known as hydrogen, is made up of nitrogen and oxygen, and that's why so-called "limescale" stains on metal-containing ones Form surfaces. If people on YouTube or other channels claim they are trained chemists with a Ph.D. but fail to recognize such false teachings, something is wrong with the universities. It is the same with self-proclaimed friends of nature who talk about prickly proteins and the dangers they emanate but do not recognize that plants and animals have prickly proteins because they develop prickles. It is abysmal.

A virus will always withdraw the host cell's glycoprotein that the virus cannot produce but needs to multiply. The existence of life as we know it is due to the bacteriophages and macrophages. Bacteriophages are always non-enveloped viruses. Macrophages, however, always enveloped viruses. Without unenveloped viruses, animals and plants would have no spines, humans would have no hair, and certain animals would have no fur. In general, the Y chromosomes arise from the bacteriophages and the X chromosomes from the macrophages. Bacteriophages invade the cell, whereas macrophages attach to the cell to prevent invasion. Bacteria, for example, need the glycoprotein thymine. It means that bacteria destroy the cell membrane (cell envelope), triggering inflammation and skin diseases. Enveloped viruses usually repair cell tissue by removing the glycoprotein cytosine from healthy cells to send them to sick cells.

However, humans are responsible for ensuring that their cells receive enough nutrients so that the enveloped viruses can also do their job. It also applies to non-enveloped viruses. Those who eat healthily pay attention to sufficient exercise and the environment will lead a long and prosperous life. The virologists are scaremongering, out of ignorance and vanity, to feel like stars for once. In my opinion, they are unsuitable for this profession. Viruses - no matter what kind - are always dependent on the glycoprotein cytosine, which means that cells can no longer divide because they start to clot, which happens when the glycoprotein thymine has withdrawn from the bacteria. As a result, cells can very well develop into cancer cells (tumors), as gas has been releasing by dead cells, which leads to the expansion of the tissue, e.g., when the tooth root dies, etc. Any virus and bacterium can cause it. Humans carry out their way of life, and behavior does not take appropriate countermeasures as already mentioned through a healthy lifestyle. The coronavirus itself is a so-called antibody, which all enveloped viruses are in general. Non-enveloped viruses such as the MERS virus are more likely to affect diseases and mutations of cells - including genetic ones - than enveloped viruses will ever be. What makes MERS viruses dangerous and what triggers measles is because they host the bacterium E. coli fimbriae.

The following picture will prove that I am right in claiming that the 5' end defines the X chromosome and the 3' end defines the Y chromosome.

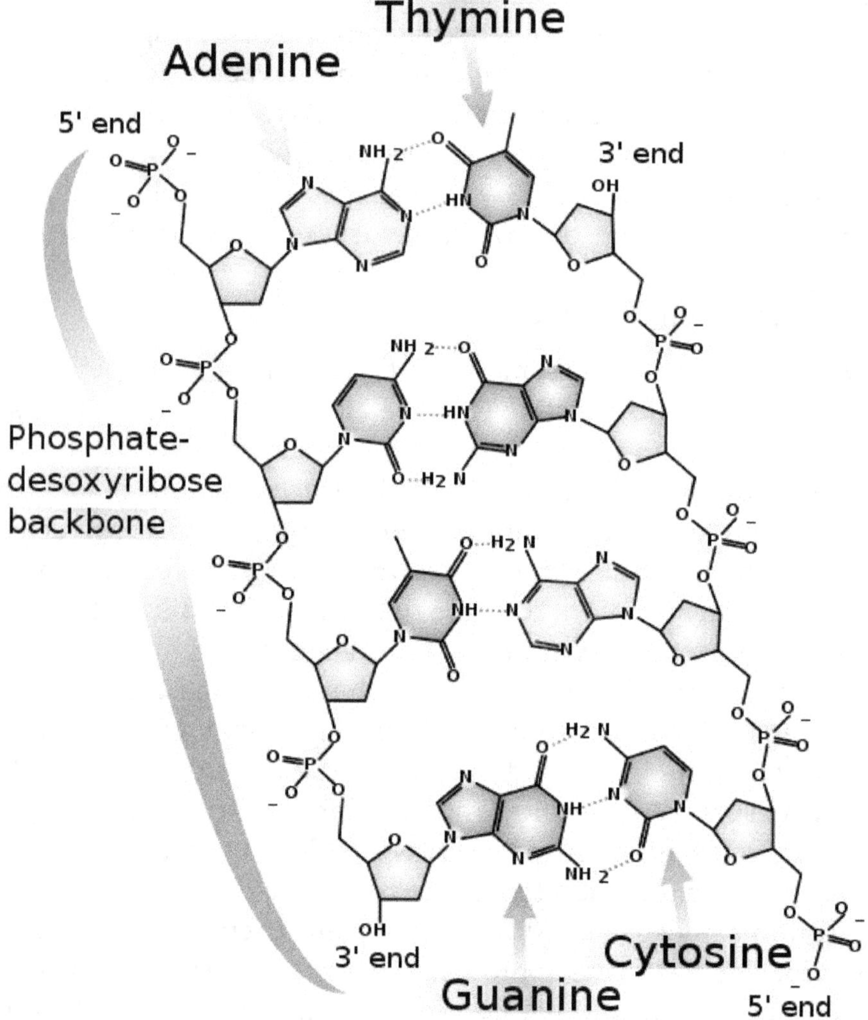

Quelle: Datei: Chemische Struktur der DNA.svg - https://de.wikipedia.org

However, in the above figure, there are errors. Namely, the adenine has blue, cytosine yellow, thymine red, guanine green, and uracil gray.

I want to give you a few examples: a mandarin consists of guanine and uracil when unripe but consists of cytosine and uracil when ripe. Or a cucumber consists of guanine and uracil in both the unripe and the mature state. Tomato consists of guanine and uracil in the unripe Form, but either cytosine and uracil or thymine and uracil when ripe. To determine which genetic structure a plant has, one only needs to analyze the starch grain accordingly since the starch grain contains the entire DNA of the plant, and everything else is then a question of development.

If HIV exists, the virus will have to trigger lactose intolerance in humans, developing into leukemia. The reason for this is that instead of cytosine, uracil has been forming as the cell nucleus. Lactose intolerance is generally a sign of diabetes, even if there is no viral infection due to a defect in the pancreas, e.g., caused by alcoholism, etc. The plague had the same effect in the Middle Ages. Since uracil has been forming in the cell instead of cytosine, sepsis developed, which caused the so-called boils. Sepsis always occurs when phagocytose is lacking and can lead to glass bones forming during embryonic development. For example, incestuous relationships favor the development of such undesirable consequences; are then phylogenetically determined, so to speak of the family descent.

Geneticists have hardly any credibility in this respect, as the cross-connections of the double helixes are not visible even under an electron microscope or transmission electron microscope, as today's technology is not yet ready. But claim that double helixes are made up of cross-connections, although they cannot see them at all, but only guess.

Telomers are the guanine product; the guanine itself is the hydrocarbons product and serves as a storage medium; see the "Bacteriophage" illustration above. On the other hand, Peplomers are the adenine product, with the adenine itself being hydrogen. What I did not know until now is that even bees need water. A linear DNA represents the X chromosomes and a supercoiled Form the Y chromosome. Reverse transcriptase turns an X chromosome into a Y chromosome, with X chromosomes being hydrophilic and Y chromosomes being amphiphilic.

Example for linear DNA:

G AT CG AC U

Example for supercoiled Form:

G AT CG AC U
U CA GC TA G

Virions were called vibrions in the times of Robert Koch and Louis Pasteur. As a result, I learned that the virions multiply through physical activity to dock on the corresponding muscle cells to supply themselves with nutrients, so that related muscle growth begins. The virions remove the cytoplasm from the cells to convert the cytosine - of which the cytoplasm consists - into guanine. You can see and feel this process because the muscle in question begins to twitch uncontrollably. However, it can also be that the virions do not withdraw the cytoplasm but the uracil, which leads to an absorption of nitrogen when more oxygen gets into the blood. It can also be the reason why white sweat stains form due to the absorption of nitrogen. The Chinese call the "new" coronavirus is nonsense. The photograph of the "new" coronavirus is missing to compare it with June Almeida's coronavirus. Virions or vibrions are spores, indicating that phagocytosis occurs in the body to assimilate microorganisms (bacteria, viruses, etc.). Everything that the body cannot use through assimilation has been excreting. Phagocytosis enables nutrients provided by foreign organisms, but what cannot be used is excreted. Virions or vibrions are the prerequisites for supplying the body with nutrients and for eliminating bacterial infections. Phagocytosis can have been using to produce mRNA vaccines, all mRNA vaccines being so-called dead vaccines. For this, I have developed the following law:

LAW OF PHAGOCYTOSIS AS A METHOD

OF MAKING MRNA VACCINES - ALSO KNOWN AS DEADLY VACCINES

dedicated to Robert Koch and Louis Pasteur

Louis Pasteur's pasteurization method is used to kill viruses and bacteria and then use them for dead vaccines, so long as they are kept in their physical state. Inactivated vaccines, i.e., mRNA vaccines, produce the phenomenon of bacterial virulence, which is also known as phagocytosis. Bacterial virulence occurs when a dormancy of viruses allows bacteria to lodge in the virus, due to which infection is symptom-free for a certain period. Pasteurization has intended to prevent bacterial virulence. At this point, I would like to bring in the "classic form" of Koch's postulates again:

1) Isolation of the pathogen in the diseased tissue.

2) Breeding in pure culture.

3) Animal reproduction of the disease using the cultures.

Extension of Koch's postulates:

4) The isolated pathogen from the diseased tissue is ground in a

mortar and pasteurized.

5) The pasteurized pathogen is then subjected to a PCR method to cultivate the corresponding viruses or bacteria.

6) With the administration of the pasteurized pathogen, the disease is temporarily triggered - at least I must assume a precautionary measure - so that a permanent and lifelong immunization occurs.

7) The finished Vaccine is called an mRNA vaccine, i.e., deadly Vaccine.

(see: Book "Robert Koch - Zentrale Texte" by Christoph Gradmann, Springer-Verlag)

8) In nature, phagocytosis is required so that, for example, parasites can develop, and predators can put themselves into a state of dormancy to transform themselves into butterflies.

Annotation:

Tuberculosis bacteria also trigger the respiratory disease. It leads to the formation of small nodules, the so-called tubercles. These tubercles then initiate metastases, leading to lung cancer or bringing people to the iron lungs in the past. The causes of lung diseases can be varied and inevitably do not have anything to do with viruses or bacteria. Respiratory diseases of the respiratory tract can occur, for example, with heart problems, poorly ventilated rooms, etc. A hundred years ago, rooms were always very poorly ventilated, especially in winter. Many working-class families spent this together in the kitchen. It was the only heated room so that all kinds of gases and smells could spread and have inhaled. But, of course, Robert Koch did not consider these things, and these possibilities are still not considered today. And that is perhaps one of the reasons why Robert Koch was unsuccessful with his tuberculin. The other reason may also be that Robert Koch was unable to identify the tuberculosis bacillus. To correctly identify a pathogen, specific properties, e.g., appearance, etc., must be present to compare these with one another, if necessary, provided that the literature allows this. Colic can also occur because of metabolic disorders, e.g., constipation, which can lead to gallstones, or long-term lack of fluids can lead to kidney stones. The tropical fever does not have to be caused by viruses or bacteria, but an acute shortage of liquid is enough to cause such symptoms.

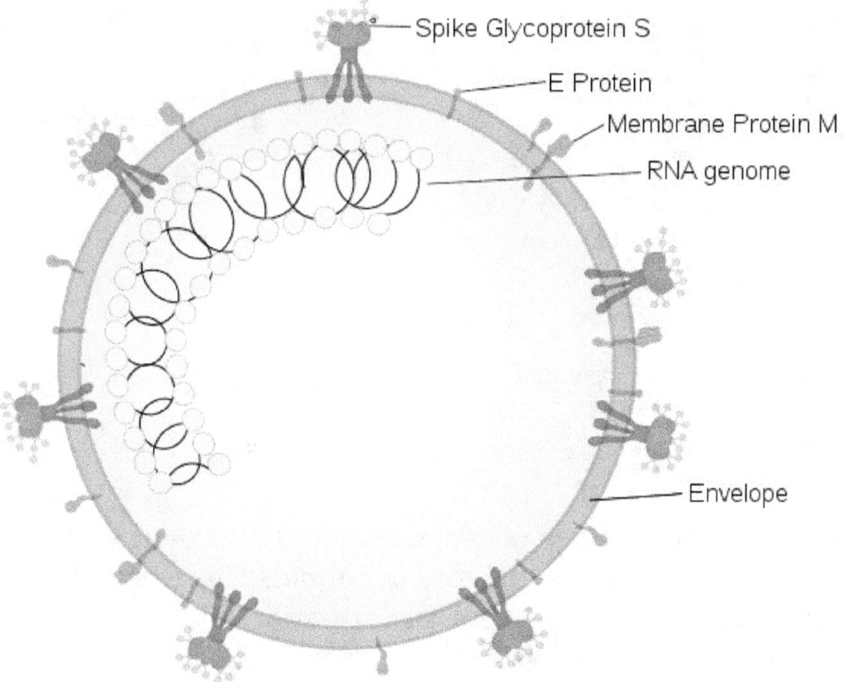

Quelle: Datei: Coronavirus virion structure.svg - https://de.wikipedia.org

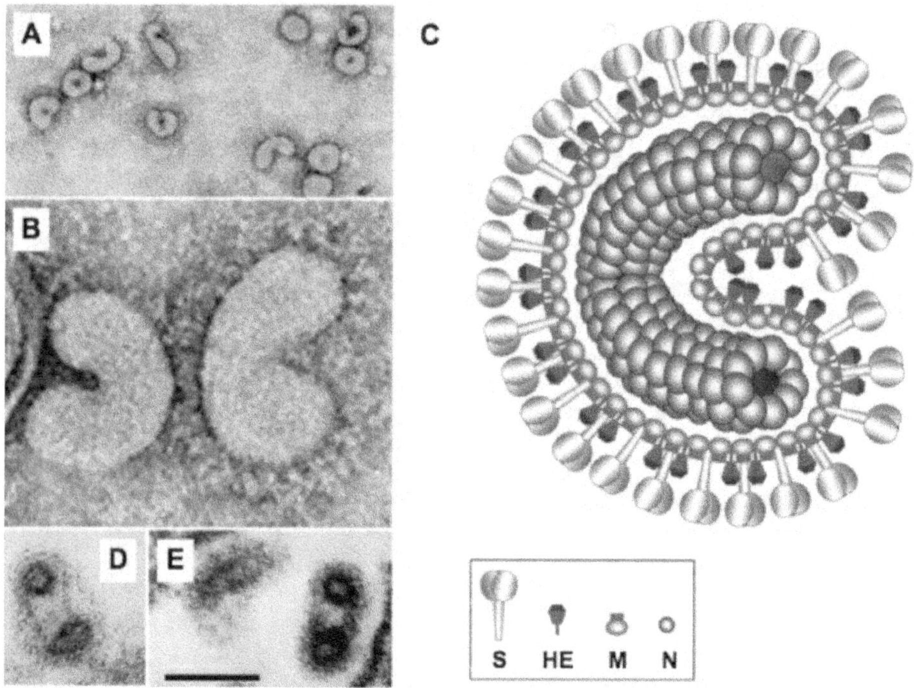

Quelle: Datei: Virion morphology of equine torovirus Berne (gen. Torovirus).png - https://de.wikipedia.org

The torovirus triggers rabies, whereby the torovirus causes the bone marrow in the spinal cord in animals. The reason for rabies is that the torovirus triggers the formation of succinic acid. The following picture shows a coronavirus with the construction of succinic acid.

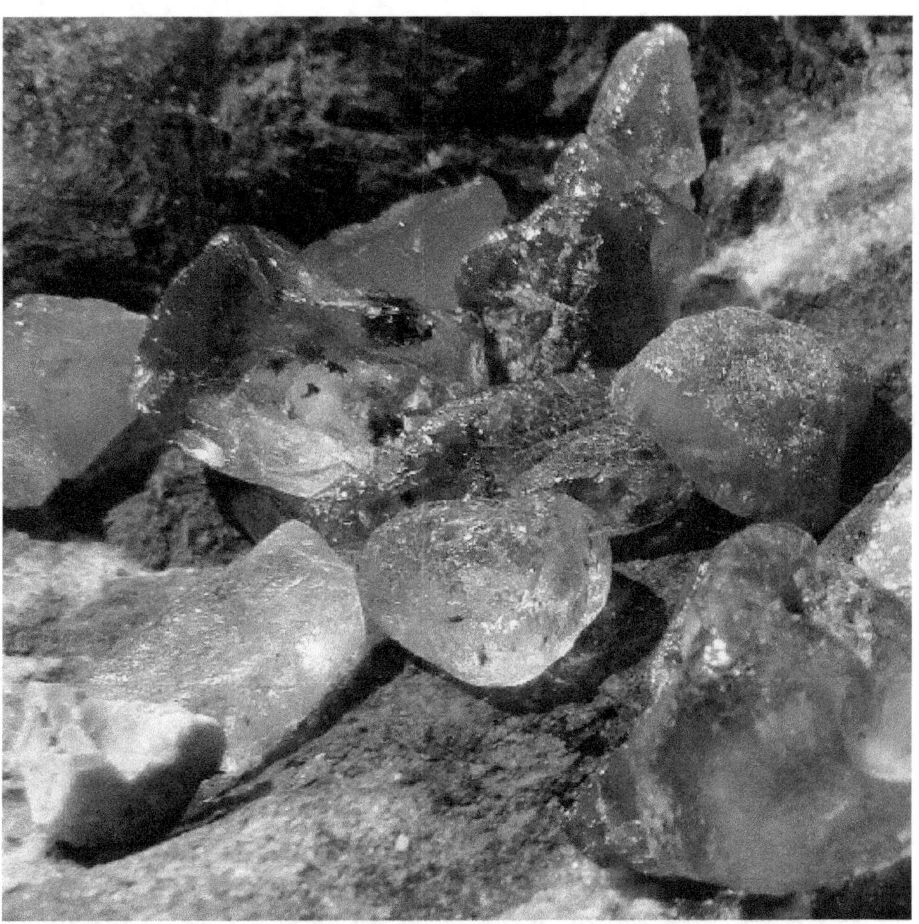

Quelle: Datei: Bernstein auf Granit.jpg - https://de.wikipedia.org

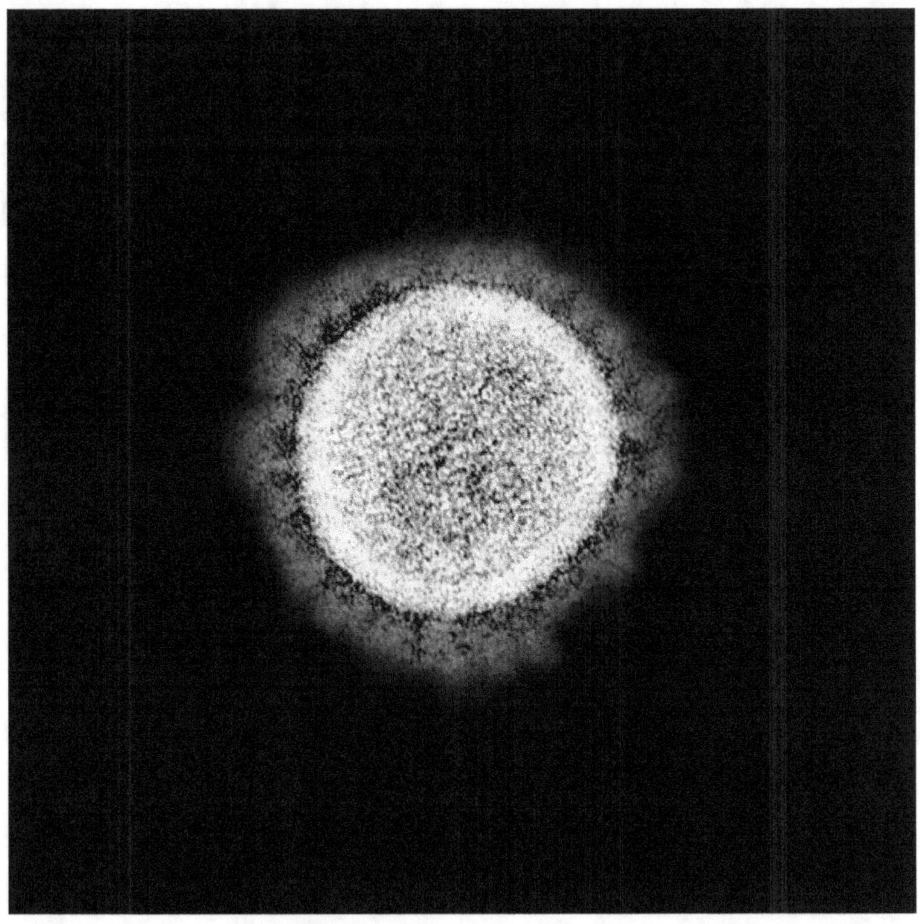

Quelle: Datei: Novel Coronavirus SARS-CoV-2 virus particles, isolated from a patient.jpg - https://de.wikipedia.org

I came across succinic acid thanks to Robert Koch. In humans, succinic acid can trigger bone diseases such as multiple sclerosis, etc., or even so-called sleeping sickness if it gets into the brain. Among other things, succinic acid serves as snake venom and can also serve as a serum against snake venom.

The coloring of viruses is irrelevant, as these would have to take on the corresponding color depending on the pigment used. Furthermore, viruses are naturally colorless due to their hydrophilicity.

The coronavirus cannot come from a bat because it has no

spines. However, since the coronavirus has spiked proteins due to its virological and genetic specificity, plants and animals must develop ranges, whereas humans build hair.

Occurring blood clots are not the reason for thrombosis, but thrombosis is the reason for clots. Thromboses then occur due to a lack of exercise, oxygen or too much uracil and nitrogen, etc. The AstraZeneca vaccine cannot be responsible for blood clots in the brain because blood clots or thromboses would have to occur in the arm through which the Vaccine has been administering. You can counteract this with blood thinners, for example.

In the documentation (only in the German language) "With viruses out of the antibiotic crisis" / Documentation Reupload / ARTE on YouTube, you can see what bacteriophages are suitable. The phages, i.e., bacteriophages and macrophages, are responsible for phagocytosis, whereby the phage therapy used is ideal for chronic inflammation. Felix d'Herelle first discovered the phages, and Georgi Eliava introduced phage therapy as the standard in Georgia. The advantage is that phage therapy - even if it is not always successful, e.g., because the wrong phages or too few phages were used - no side effects occur. The only side effect is that failure will only revive a pre-existing inflammation. Therefore, instead of the phages, I would add their lysine to the mRNA vaccines that are already available and dispense entirely with phages. However, the bacteriophages like these have shown in the documentation, but it looks more like T helper cells. In Tbilisi, Georgia, a corresponding center has set up by Georgi Eliava and Soviet support. The phages are taken either orally as a liquid or directly into the inflamed area. The isolation of the phages was carried out by Felix d'Herelle as follows: 1) A transparent nutrient medium was inoculated with bacteria; after hours, it becomes cloudy. 2) The bacteria in the gray nutrient medium were infected with phages and died, producing new phages; the nutrient medium cleared up again. 3) The nutrient medium was filtered through a porcelain filter that kept bacteria and other larger objects out; the small phages could pass through the filter. 4) I would add succinic acid to the existing mRNA vaccine so that the human DNA would have been expanding to include the element succinic acid so that the DNA would then read as follows. I would then call it lysine, a simplified term for an everyday language that leads to a corresponding awareness: GATCB for bacteriophages and GACUB for macrophages.

The current COVID rapid test cannot work at all. It does not recognize any difference between which illness or previous illness led to antibodies' formation. So, the question is still open, whether the antibodies are always the same or different in different diseases.

When viruses multiply, a linear DNA chain has created naturally. This linear DNA chain - read: macrophage - feeds on bacteria to splits the linear DNA chain. This process repeats itself repeatedly, which is known as a biological polymerase chain reaction. A reverse transcriptase, on the other hand, occurs when bacteria feed on viruses. Still, they cannot divide themselves but absorb a virus, generate the virus themselves, and then excrete them again. Thus, bacteria are the producers and spreaders of viruses.

Phages attach themselves to a bacterium and then form what is known as the lipid membrane envelope, which makes bacteria behave like viruses. Macrophages have also been using when phages attach themselves to a bacterium because they have a lipid membrane cover. The corresponding bacterium begins to eat other bacteria through the phages.

Phages are all the same so that only the amount of phages administered determines a therapy's success. According to Dr. Mzia Kutateladze and Felix d'Herelle, they can convert the wastewater into clean drinking water. But if that were the case, then in the past, there should not be any dysentery diseases that lead to renal colic, etc., when consuming unclean water.

ACT TO REVIEW

THE MRNA VACCINES EFFECTIVENESS

Özlem Türeci and Dr. Mzia Kutateladze

1) You need six rapid PCR tests in advance. 2) The first quick PCR test is performed with the mRNA vaccine only and without a human component, and the device must respond with a positive test. 3) To avoid falsifications, another rapid PCR test has been using. A sample of coronaviruses, which may only be in the water due to the culture's purity, has been adding. It must also react with a positive value. 4) The third rapid PCR test is only one human component added, but the value here must be negative. 5) The fourth quick PCR test is then with a human element. Simultaneously, the mRNA vaccine has been adding to this human component. The value must also be harmful because the human part has a different RNA structure than the coronavirus. Both together result in a completely different RNA structure, which is the reason for a negative value. 6) The fifth rapid PCR test is then carried out on vaccinated test subjects. The device must display a positive value because the human component will have adapted to the coronavirus's RNA structure so that antibodies must have formed accordingly. Still, what does take a while and why in 5) the value must be negative. 7) The sixth rapid PCR test is used to check the duration of the mRNA vaccines' effectiveness by keeping the values positive in weekly rapid PCR tests so long as the Vaccine's effectiveness continues. 8) These postulates can be applied to all diseases when testing mRNA vaccines or even classic vaccines in the rapid process. But also, the func-

tionality of the device used for the PCR tests.

DNA Image:

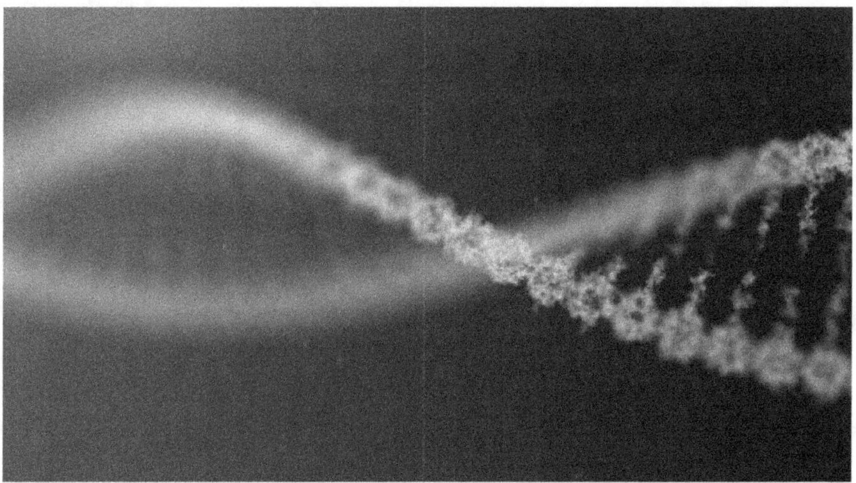

Bacterial virulence - in other words: phagocytosis - ensures that any Vaccine loses its effectiveness, as all viruses and bacteria will adapt accordingly and even naturally must. Viruses can very well multiply, even without a host, provided they have a lipid membrane cover. As the picture above shows, the glycoproteins' arrangement in an RNA and DNA sequence is purely random since the picture above does not show where glycoprotein has located. It is not possible to develop a suitable primer that recognizes the correct sequence or the individual glycoproteins. Mechanically, i.e., from a machine perspective, if the primer matches the respective RNA or DNA sequence, a harmonic oscillation will have to be generated, but if there are differences, a damped oscillation. For this reason, I allow myself to reinterpret the previous Koch postulates accordingly:

DROSTEN POSTULATES

Prof. Christian Drosten

1) Isolate the diseased tissue and place it in a test tube, possibly with a distillate.

2) Heating the contents of point 1).

3) Then centrifuge.

4) Use the pipette to take a correspondingly large amount for the PCR test.

5) Perform the PCR test.

6) As a result, the necessary mRNA is obtained, which should first have used in a corresponding laboratory test.

7) When phagocytosis begins, which also takes place without phages, lysosomes must form, from which the antibodies then emerge, which are also known as DNA viruses - i.e., coronaviruses.

As a critic, I would like to point out that evolutionary biology has left out virology and epidemiology. But without this, we will never really learn anything about nature. Infections are symptom-free because viruses turn into spores, initially in a dormancy state (hibernation). Occurring phagocytosis occurs depending on the living and environmental conditions, deciding which species-specific properties plants, animals, and humans will have. As a result, phages prevent epidemics and pandemics from occurring and prevent genetic predispositions from occurring. The placenta also takes over the task of phagocytosis.

Dr. Clemens G. Arvay is quite right when he claims that individual nucleotides (or glycoproteins or amino acid bases, etc.) are in the same place or, to put it more cautiously, can be. And at this point, I would like to provide the necessary evidence and explain the mechanical background. And record this in a corresponding law.

LAW OF DNA

Dr. Clemens G. Arvay and Dr. med. Umes Arunagirinathan

and Dr. Mike Hansen

1) Linking matrices:

$$\begin{bmatrix} 1 & \rightarrow & 0 \\ \uparrow & \ddots & \downarrow \\ 0 & \leftarrow & 1 \end{bmatrix}\begin{bmatrix} 0 & \rightarrow & 1 \\ \uparrow & \ddots & \downarrow \\ 1 & \leftarrow & 0 \end{bmatrix}\begin{bmatrix} 1 & \rightarrow & 0 \\ \uparrow & \ddots & \downarrow \\ 0 & \leftarrow & 1 \end{bmatrix}$$

Explanation:

As you can inevitably see within this representation, the matrix link means that link or genetics always shows the same value twice in a row, e.g., zero or one. As a geneticist, you then know that cells have divided at this point. The zero value stands for the cell membrane, i.e., reverse transcriptase, and the one value for the cell nucleus, i.e., protease. So, it means a link and the arrow directions should again show the angular frequency in computer science.

Now we are going to apply this knowledge in genetics. It then results in the following picture, the above RNA strand representing the positive polarity and the lower RNA strand representing the negative polarity.

So, we get the following scheme:

Y chromosome

$$\begin{bmatrix} G & \rightarrow & C \\ \uparrow & \ddots & \downarrow \\ A & \leftarrow & T \end{bmatrix}\begin{bmatrix} A & \rightarrow & G \\ \uparrow & \ddots & \downarrow \\ T & \leftarrow & C \end{bmatrix}\begin{bmatrix} T & \rightarrow & A \\ \uparrow & \ddots & \downarrow \\ C & \leftarrow & G \end{bmatrix}\begin{bmatrix} C & \rightarrow & T \\ \uparrow & \ddots & \downarrow \\ G & \leftarrow & A \end{bmatrix}\begin{bmatrix} G & \rightarrow & C \\ \uparrow & \ddots & \downarrow \\ A & \leftarrow & T \end{bmatrix}$$

X chromosome

$$\begin{bmatrix} G & \rightarrow & C \\ \uparrow & \ddots & \downarrow \\ A & \leftarrow & U \end{bmatrix}\begin{bmatrix} A & \rightarrow & G \\ \uparrow & \ddots & \downarrow \\ U & \leftarrow & C \end{bmatrix}\begin{bmatrix} U & \rightarrow & A \\ \uparrow & \ddots & \downarrow \\ C & \leftarrow & G \end{bmatrix}\begin{bmatrix} C & \rightarrow & U \\ \uparrow & \ddots & \downarrow \\ G & \leftarrow & A \end{bmatrix}$$

$$\begin{bmatrix} G & \rightarrow & C \\ \uparrow & \ddots & \downarrow \\ A & \leftarrow & U \end{bmatrix}$$

What annoys me is that I have been teaching better again because now, after being confronted with the scheme, I realized that Dr. Clemens G. Arvay's finding might be wrong. In this present scheme, the reverse transcriptase takes place in the lower area, but in the upper row, it is not clear what change is taking place, other than that one element is converted into another and that in an endless cycle. Suppose such complications (incommensurability) occur in the interpretations of an individual like me. What will it be like for geneticists who must trust their arrangements made under machines' guidance? However, this law reveals the future when it comes to cloning humans or animals. You can only use his semen when cloning a man and only the egg cell when cloning a woman. You would have cloned a person who looks like the original but has an entirely different character in both cases.

It should also have been noting that the RNA or DNA of a virus cannot be detected using the PCR method because the amount of protein required by a virus is far too low and because an RNA strand cannot have been isolating.

So far, it should be enough, and the topic "Corona" and genetics will end for the time being as the risk of endless repetitions and contradictions increases. Ultimately, the method used decides what is correct or not. Still, the disadvantage is that the method used is always dependent on the premises of the interpretations that always precede the forms. As an outsider, you should ask yourself the following questions: 1) Why is such a play performed right now, especially with this virus? 2) How does a virologist recognize the danger of viruses and bacteria, or what criteria decide which virus or bacillus is more dangerous than others? Of course, I got information from Wikipedia and various books such as "Corona Vaccines - Rescue or Risk" by Clemens G. Arvay or "Robert Koch - Central Texts" by Christoph Gradmann published in German. It does not matter whether I am right or not in the end; the only thing that counts is that there are possibilities of contradictions because too much vanity through money, power, prestige, and property robs scientists' minds. Thus, all the science discredits when pharmaceutical interest groups' satisfaction is more important than anything else. In the past, Robert Koch himself tried to enrich himself by selling tuberculin - at least this can have been reading in the book: "Robert Koch - Zentrale Texte" - and has done himself no favors and even the sleeping sickness that was in from today's point of view, I can call Egypt supposed to have existed, even if I was able to point out a possibility of such a disease occurring. But that is how I change myself and my views over the days, weeks, and months, but I am always ready to be taught better. Because everything I write, I write primarily for myself. The booking form has only been used for better readability, making myself happy, and dealing with current or past things helps me think emotionally and improve. But, if I have learned one thing, it is not to believe every Bullshit in books, YouTube videos, or the news. How many more books and theories should have been writing when everything is as straightforward as claimed? Why didn't virologists and epidemiologists deal with SARS back in 2003? If they do, why are they all unable to apply what they have learned now? There are always questions, but no one gets plausible, consistent, and understandable answers, and I am not the only one who feels this way. Still, I try to deal seriously with the problem, which journalists, etc., did not have necessarily held.

All the virologists and epidemiologists I have considered have one thing in common: they have not made any significant contribution through their past work—none of the virologists and epidemiologists appearing on TV or otherwise have had any track record. Most doctoral dissertations are not an authentic doctoral thesis but only school presentations since today's doctoral thesis only aims to copy and present what is in books. The only services virologists have so far have been to examine fungal spores and nothing else at all. None of them did research. Otherwise, they would be free of contradicting statements since competitive behavior is also spreading in science because everyone wants the most significant possible recognition. And even Chancellor Merkel failed in her physics studies, and for this reason and in this area has nothing to show, let alone achieved a degree, because to do this, she would first have had to come up with her theories that bring to light the insights and perspectives that still exist not gave or gives. And that is no different from a doctoral thesis. For example, I only need to look at Albert Einstein's work, bringing new insights and light perspectives. Could Albert Einstein achieve a doctorate because a doctoral thesis must go beyond the already known knowledge state?

Today, the public and even Wikipedia do not know what the SARS-CoV-1 virus looks like objectively. How do the Chinese know whether the SARS-CoV-2 virus is another coronavirus since all viruses with a lipid membrane envelope can replicate themselves, even without a host?

In retrospect, we look at ourselves and our surroundings much more closely. We find that excessive fear or fear drives us to turn our lives into a business that will bring us fame, money, and what we mean by power. So false pride becomes the flip side of a worthless coin. We must be first-class people to cover up our deep-seated shortcomings. True ambition is different from what we imagine. Genuine aspiration is the desire to live and experience a meaningful life by the grace of God.

At this point, I would like to add a few additional ideas:

The antigen test for laypeople that I ordered is unfortunately invalid. The reason is that, in the case of a negative corona value, both C (cytosine) and T (thymine) must have been displaying and not just C (cytosine) as in the case of the antigen test. So, the color is not decisive. But I claim that this test can have improved. So, Beijing Hotgen Biotech Co., Ltd. will at some point discover that with honesty, you not only get further in life but can also earn money just as successfully, even outside of pandemics.

So, I will make a few suggestions to improve it:

1) The sample extraction buffer should consist of G, A, T, C, but C and T may be sufficient. At least it is my recommendation that both have tried.

2) As before, take a swab from both nostrils.

3) For a negative corona value, the test must show C and T, regardless of whether the sample extraction buffer consists of G, A, T, C, or just C and T.

4) Reason:

As soon as the cytosine inside a cell has used up, no more cell division occurs. Instead, uracil has formed. Uracil triggers a mutation of cells, so that cell division has been preventing.

In between, I once again criticize the geneticists: "Why are geneticists not even able to answer the question of why there is uracil instead of thymine in RNA and thymine in DNA instead of uracil?"

REVERSE TRANSCRIPTASE LAW

Dr. med. Umes Arunagirinathan and Dr. Mike Hansen

A reverse transcriptase corrects errors in individual bases, with a base excision repair by a tRNA, leading to menstruation beginning or have been creating. The tRNA removes damaged amino base sequences from the catalytic center, releasing them either through menstruation or the intestine. It triggers a DNA polymerase that links to the DNA strand by forming a new amino acid sequence. Only a single base has been replacing in an RNA case, known as a short patch repair. On the other hand, in DNA, between 2-20 nucleotides are replaced, known as a long patch repair. The DNA ligase is based on an AP endonuclease by a tRNA responsible for an RNA that only a single base has replaced. The DNA glycosylase, on the other hand, is based on an AP endonuclease by an mRNA, which returns 2-20 nucleotides in DNA. Namely, bacteriophages are those microorganisms that produce a tRNA, and macrophages are those microorganisms that produce an mRNA. If the lay antigen test follows this law, then validation can take place. The reason is that all viruses that have a lipid membrane envelope are coronaviruses and those microorganisms that can remove cytosine from the cells to multiply. It means that uracil is always used instead of thymine in a DNA strand, and thymine is absent.

I want to make a small note concerning the law, namely that I have adapted the original text in Wikipedia accordingly to this law.

I want to add a few more details about the so-called glycoproteins, which are composed as follows:

1) Glucose is made up of cytosine and uracil.

2) Glycose is made up of thymine, cytosine, and uracil.

3) Glycogen consists of adenine, cytosine, and uracil.

4) Guanine creates a bitter taste.

Vector vaccines must constantly be introduced directly into the inflamed tissue of the relevant organ. Otherwise, vector vaccines wrongly have this name. The rapid corona test for laypeople from Beijing Hotgen Biotech Co., Ltd. showed me that too long a transport route, e.g., from the shoulder to the lungs, causes a material to change this time. Also, I consider it nonsense to expect a vaccine injected into the shoulder and transported to the lungs to stimulate cell division. A vaccine gets to where it is needed; bacteriophages or macrophages have required. Bacteriophages have based on a tRNA, an animal RNA, so to speak, whereas macrophages have been finding on an mRNA, i.e., plant RNA. I claim, however, that animal, vegetable, and human RNA or DNA do not have any characteristics that the human mind can distinguish - apart from the difference between RNA and DNA, which consists in the fact that an RNA instead of thymine uracil and a DNA instead of uracil owns thymine. With a PCR test, not even with the so-called antigen tests for laypeople, e.g., from Beijing Hotgen Biotech Co., Ltd., or others, the infectivity that viruses produce or have intended to create can never be detected. Therefore, all these tests should only give positive or negative or invalid results in the worst case.

The whole supposed knowledge about genomes - no matter what kind - is based on pure speculation and is why genetics has not succeeded in combating genetic dispositions - regardless of mental or physical disabilities - even after decades. If I were wrong, the United States should have the world monopoly on gene-based vaccines, but instead, it has nothing.

A reverse transcriptase depends on the mRNA used in each case to obtain exact copies of the original RNA strand or the outgoing RNA strand so that an unadulterated DNA has been revealing here. Finally, I will show some examples of what happens when the mRNA does not match the RNA and what transformation means in virology and genetics.

Examples:

I always use the same RNA to ensure better traceability for all examples, which applies to the DNA.

It must have been considering that with an RNA, only one nucleotide is ever exchanged or converted.

$$1)\ RNA{:=}A \quad B{\rightarrow}mRNA{:=}C \quad C{\rightarrow}new\ RNA{:=} \\ A \quad C \quad C$$

The nucleotide with the letter B was converted into a nucleotide with the letter C by the mRNA.

$$2)\ RNA{:=}G \quad AC \quad U{\rightarrow}mRNA{:=}G \quad AC \quad T{\rightarrow}$$

$$new\ RNA{:=}G \quad AC \quad G \quad AC \quad T$$

In the present case, the nucleotide with the letter U was converted into a nucleotide with the letter G by simply joining two different RNAs by causing conversion of the nucleotide U into a nucleotide G.

etc.

$$3)\ DNA{:=}\begin{matrix} G & C \\ A & T \end{matrix} \rightarrow mRNA{:=}\begin{matrix} G & C \\ A & T \end{matrix} \rightarrow$$

$$new\ DNA{:=}\begin{matrix} G & G & C \\ A & A & T \end{matrix}$$

In the following example, the one new DNA has been achieving by converting nucleotide C into G from the previous DNA in the above part and nucleotide T into A in the lower leg.

$$etc.$$

Now I would like to show the following matrix connection:

$$Matrix\ A{:=}\begin{matrix} 1 & 0 \\ 0 & 1 \end{matrix} \rightarrow Matrix\ B{:=}\begin{matrix} 1 & 0 \\ 0 & 1 \end{matrix} \rightarrow$$

$$new\ Matix\ C{:=}\begin{matrix} 1 & 1 & 0 \\ 0 & 0 & 1 \end{matrix}$$

By linking matrix, A with matrix B, the new matrix C has been obtaining, because the number 0 in the upper part has converted into a 1, and in the lower leg, the number 1 has transformed into a 0.

$$etc.$$

After a while, I found out that the German language is the key to deciphering genetics. Therefore, I would also like to bring a new law on complete genetic analysis into play.

LANA SANDLER LAW

This law proves that human and animal cloning is dependent on male and female DNA. Before, I thought that cloning would be possible without this dependency, but it is not. This dependency does not exist in the plant world. According to Dr. Clemens G. Arvay, the amino acid bases are always opposite in female DNA. Namely, guanine has been comparing to cytosine and adenine compared to thymine, so guanine and cytosine form a base pair, and adenine and thymine together result in a base pair.

1) **<u>Y chromosome (tRNA):</u>**

G AT CC TA G
C TA GG AT C

G AT CC TA G
C TA GG AT C

When the same amino acid appears next to each other twice, cells divide at the point.

2) **<u>X chromosome (mRNA):</u>**

G AU CC UA G
C UA GG AU C

G AU CC UA G
C UA GG AU C

At this point, the same amino acid appears twice next to each other, cell division, and through this same amino acid.

Guanine does not always base pair with cytosine because an unripe tomato has green skin, converted to thymine by exposure to UV light, which means that guanine has converted to thymine by doing this also forms a standard base pair. It also applies to all types of fruit and vegetables, to the entire plant world.

It should also have noted here that the yellowing of previously green leaves has to do with the fact that the guanine has been using up because the season changes. It has nothing to do with a virus but simply with the laws of thermodynamics. Human and animal cells also depend on the presence of a specific solar activity to guarantee an ATP synthase. Otherwise, as with falling temperature, the conversion process slows down so that there is a deficiency, and when the temperature reaches a critical moment, cells die. And the same thing happens with plants. Permutations also occur in genetics since a genome never has the same sequence of X and Y chromosomes. Therein lies the difference between humans, animals, and plants, even if they belong to the same species.

Fungal mRNA has been obtaining from the spores from which coronaviruses develop. It also applies to the male sperm cell in humans and animals. The egg cell in females and coronaviruses in fruit and vegetables is obtained from the starch grains located inside kernels.

All rapid antigen tests are flawed as they do not distinguish between the corona and human genomes. Therefore, it should be positive or negative, but my experience shows that the rapid test always shows negative and never a positive value. In the event of an infection - regardless of which virus - the quick antigen test should only display T since an infection suggests that no cells are dividing due to a lack of C. No matter which RNA or DNA sequence a virus has, this has no effect on genetics, but only in the case of embryonic development, but not in those already born. A virus would have to grow into a parasite in the body to cause damage. Unfortunately, there is no distinction between inorganic and organic chemistry since only inorganic substances disturb the immune system or the vegetative nervous system. Still, organic substances stimulate the immune system or the vegetative and central nervous systems. The geneticists also have no idea of the chemical composition of the respective amino acid, so I must correct that again. The reason is that hydrogen is naturally composed of oxygen and nitrogen, with oxygen forming the atomic shell and nitrogen being the atomic nucleus.

All rapid antigen tests have a saline solution as sample extraction buffer, and that is also the reason why this test always only shows C. If someone like me has already manipulated this test several times to find out whether it recognizes a difference, this test should at least deliver an invalid test result, as it is in the description.

Guanine is an element that can cause severe damage, but it is also essential for life. From a safety sheet for handling guanine $\geq$ 97%, I learned the following: 1) Guanine causes skin irritation (code: H315), severe eye irritation (code: H319), and irritation of the respiratory tract (code: H335). Besides, guanine is flammable and has a light beige color.

Inflammation is because there is an excess of guanine and uracil, e.g., UUGGG, etc. Cancer arises from this fact. Uracil consists of NH_2O and says that the atomic shell consists of nitrogen and the core of helium and oxygen. Guanine, on the other hand, consists of NH_3CO_2. NH_3CO_2 stands for hydrocarbon. Adenine, on the other hand, consists of $NHCO_2$. Thymine has made up of H_3CO_2. Finally, the cytosine consists of HCO_2.

MAI THI LEIENDECKER - HEISENBERG'S LAW OF UNCERTAINTY

IN CHEMISTRY FOR DETERMINING THE PHYSICAL PROPERTIES OF CHEMICAL ELEMENTS

by Mai Thi Leiendecker und Clemens G. Arvay

In chemistry, quantum electrodynamic processes have been describing differently than in quantum physics. For example, although oxygen provides electrons, electrons do not generate a chemical element because chemical elements create frictional losses that do not exist in quantum physics. But Heisenberg's uncertainty principle is the law. Its strength is that it considers both the weak interaction that leads to chemical processes due to frictional losses and the strong interaction that leads to physical methods due to a lack of frictional losses.

For this reason, oxygen has created when an electron causes the helium to scatter. When a proton scatters oxygen, helium has made. But a nitrogen atom meets an electron, and hydrogen has been creating. When a carbon atom meets an electron, we get hydrocarbon. If, on the other hand, a carbon meets a proton, then we get phosphorus.

The Heisenberg uncertainty principle generally states the following: "If something eludes direct observability, the Heisenberg uncertainty principle must have determined by determining the wavelength of the respective elementary particles, chemical elements, etc."

Their fast antigen test should be positive for all those already vaccinated, but it will not. The whole pandemic is a hoax. Mr. Wodarg had already been right back then when it came to bird flu and swine flu. I recommend the Toni Bartl channel on YouTube video: "History repeats itself ... Almost 1: 1" (unfortunately only in German). I can only recommend the other videos; how other poorer countries feel about Germany. Make fun of politicians.

The corona pandemic is a farce since the entire World Health Organization is an interest-led pharmaceutical company. Either the members work in a pharmaceutical company themselves or are rewarded by pharmaceutical companies. They receive a corresponding position or own a laboratory for vaccine production, which is 100% financed by pharmaceutical companies. The World Health Organization has lowered the pandemic level in favor of pharmaceutical companies so that a pandemic can now have declared much faster. The federal government already has a contract - at least I could find out in the video - with ClaxoKline-Smith.

Bill Gates also earns money from the corona pandemic because he puts his fortune in pharmaceutical companies, and his "generosity" towards poorer countries is only a means to an end. Once a businessman, always a businessman.

And the more videos I watch on the subject of "pandemics" from the past, the greater the certainty that the whole corona pandemic is rubbish led by pharmaceutical companies. That is also why Bill Gates always poses as a prophet, making money from it himself.

All statistics on the corona pandemic have been manipulating because 1) the rapid antigen test always delivers a negative value and 2) the rapid antigen test should always show positive for people who have already been vaccinating since they must have formed the corresponding antibodies, which but the Fast Antigen Test will not show.

Also, the virologists are all liars. When they claim that they can determine the DNA of a virus since genetics has not yet developed that far since geneticists are still not able to decide on the individual properties of a human DNA when it comes to determining which part of the DNA is responsible for the development of eyes or brain or hair color or the like. What geneticists cannot do; virologists cannot do for a long time.

At this point, I would like to state the reason why I am sure of my case that all vaccines are the only placebo, because the vaccine 1) is colorless, which should never be when mixed appropriately, and 2) because nobody knows how it is "Coronavirus" really builds up genetically when it comes to analyzing the corresponding amino acid sequences.

To determine how the individual amino acids have built up, one only needs to consult the flora. If guanine is naturally green, the element will retain its green color even in its liquid state. If you want to know how guanine has made up, you only need to remove a wafer-thin slice of cucumber, for example, put it in a test tube, heat it slightly, and analyze the resulting liquid using a mass spectrometer. You only need red tomatoes for thymine, as there are also yellow ones, and they carry out this process. It is also done with cytosine, for example, by taking mandarins, and for the uracil, you only need to use a lemon and squeeze it. You take a blue plum for the adenine and cut a wafer-thin slice from the skin, but only from the skin. It should have been noting that no additives have been using in order not to falsify the result.

Because of this, the Vaccine would have to be a completely different color than it currently has. I claim that the Vaccine used is just a saline solution like the one used in hospitals and nothing else. What is objectionable about the medical professionals is that they cannot differentiate between people with coronary heart disease and lung problems because people with severe lung problems always give off bloody sputum.

Experts like Mr. Wodarg always make a mistake, which the so-called "lateral thinkers" in Germany also make; they never support their assumptions by uncovering the contradictions within the generally accepted doctrines.

But that is what I want to show with this book. Because if you are going to criticize, you should first familiarize yourself with the subject matter of said sciences. Otherwise, any criticism is unfounded and will have been treating accordingly. The contradictions that arise with a layman like me are completely normal due to the personal lack of experience and learning process. I am no better or worse off than the students at a university, provided they study this subject. But virologists who claim to know more about genetics than the geneticists themselves are pure presumptions on the virologists' part.

The whole corona pandemic is due to indoctrination on the part of the dependency of the World Health Organization on the interest representatives of the pharmaceutical industry because the World Health Organization can only receive money from the interest representatives of the pharmaceutical industry on a large scale because for all other interest groups of the industry or like that, there is no reason to invest in the World Health Organization.

I would also like to note that the safety sheet, which I have already mentioned, can confidently be thrown in the bin, as any amino acid can pose a risk to skin, eyes, etc., as the term acid already implies this risk.

I would also like to point out that most of the supposed knowledge is based only on pure speculation because concepts are just human achievements. Therefore, we will always be dependent on constantly questioning the fundamental ability anew, which is a task of every new generation when every generation wants to allow itself a certain amount of progress.

To come back to Clemens G. Arvay's claim that guanine is always opposite cytosine is nonsense. For example, an unripe tomato has green skin but red skin when ripe, which suggests thymine. Of course, there are also yellow tomatoes because of the conversion of guanine into cytosine. But any element can have transformed into any other feature required by the body.

Viruses trigger phagocytosis in every organism, which in biology is called metabolism, so that an excess of guanine, for example, is converted into another substance required by the body. In physics, one, therefore, speaks of thermodynamics.

To come back to the color spectrum of the amino acids: guanine is always green by nature, thymine is always red by nature, cytosine is always yellow by nature, adenine is naturally blue, and uracil is naturally colorless or white.

I have just tasted the liquid that serves as the sample extraction buffer for the rapid antigen test. It tasted bitter at first but salty afterward. That tells me personally that it is just a saline solution.

Even if the corona pandemic is fake, it has its good sides. The pandemic shows how people are in times of crisis and that the pharmaceutical industry is so bold that it gives people vaccines that have no effects. If I am honest, I can hardly contain myself laughing. I have never learned so much during this time, and that is why I can laugh my ass off.

Clemens G. Arvay is quite right when he claims in his book that guanine is the opposite cytosine, and thymine is the opposite adenine. Accordingly, I will again formulate a law, for every lawfulness is the essentiality of ethics, and practicality is the substantiality of morality.

What I would like to add at this point is that the contradictions arise because 1) the information in the works of literature and official databases are contradictory and 2) is due to the learning process to be able to develop a corresponding awareness of this matter and to show it that everything is a matter of interpretation. Targeted experimentation is never possible without theories. Otherwise, you make yourself dependent on chance.

Suppose you are right of the opinion that I do not have the necessary skills for this matter. In that case, I can get a correspondingly differentiated picture of science, and the virologists and geneticists are guaranteed not to have more but somewhat less knowledge than me.

"Coronaviruses" are the carriers of the so-called tRNA and are of animal nature. On the other hand, T helper cells are the carriers of the mRNA and are of a plant nature.

The anthrax pathogen triggers the plague and has been starting by the pathogen Yersinia pestis. The anthrax pathogen, i.e., the plague pathogen Yersinia pestis, causes blood poisoning, which leads to sepsis, and so-called boils form.

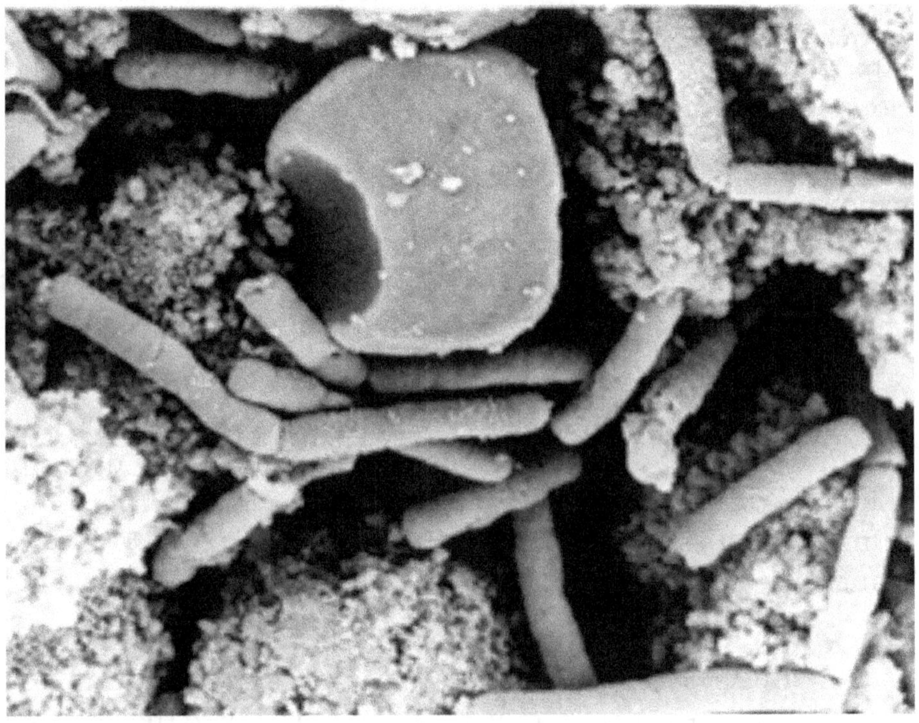

Quelle: Datei: AnthraxBacteria.jpg - https://de.wikipedia.org

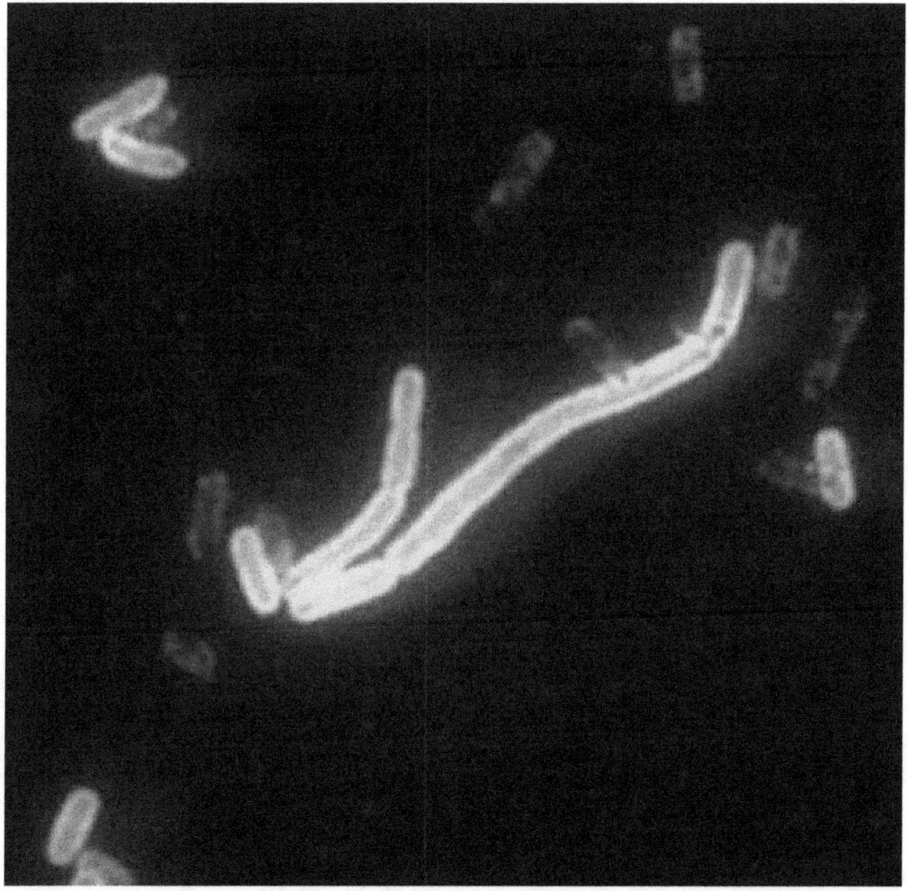

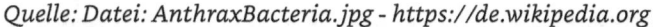

Quelle: Datei: PHIL 1918 lores Floureszenz Yersinia.jpg - https://de.wikipedia.org

The plague pathogen is also the trigger for cattle swine flu, bird flu, and foot-and-mouth disease. The reason is that the soil is contaminated by the animal dunk, as there are already bacteria in the animal dunk that can trigger the plague. In addition, the animal dunk contains ammonium nitrate - i.e., ammonia - which makes animals, people, and plants sick.

Bad animal husbandry makes animals sick, and the meat is then of poor quality. And what makes animals sick makes people sick even more.

Now the proof for the correctness of the statement of Clemens G.

Arvay regarding genetics:

CLEMENS G. ARVAY LAW

Dr. Clemens G. Arvay

$$\begin{matrix} G & AT & CC & TA & G \\ C & TA & GG & AT & C \end{matrix}$$

Definitions: **tRNA**

1) X chromosome:

$$\begin{matrix} A & T \\ T & A \end{matrix} \text{und} \begin{matrix} T & A \\ A & T \end{matrix}$$

2) Y chromosome:

$$\begin{matrix} C & C \\ G & G \end{matrix}$$

The Y chromosome is composed as follows, and this part is responsible for reverse transcriptase:

$$\begin{bmatrix} C & \leftrightarrow & C \\ \downarrow & \nearrow & \vdots \\ G & \cdots & \blacksquare \end{bmatrix} und \begin{bmatrix} C & \cdots & \blacksquare \\ \uparrow & \searrow & \vdots \\ G & \leftrightarrow & G \end{bmatrix}$$

A Y chromosome with the corresponding reverse transcriptase property can have recognized by the fact that a triangle can have formed.

In an X chromosome, identical elements must be diagonally opposite each other.

There must be a Y chromosome between two X chromosomes or an X chromosome between two Y chromosomes to speak of protein synthesis or protein biosynthesis.

3) <u>genetic disposition:</u>

$$5'|U \quad U \quad GG \quad G|3'$$
$$3'|G \quad G \quad GU \quad U|5'$$

This example clearly shows that the X chromosome is missing between the two Y chromosomes, and a deformity arises. This effect creates a mutation in viruses and bacteria.

4) <u>Difference between telomere formation and peplomer formation:</u>

So-called bacteriophages, which are responsible for the creation of the viral RNA, arise from the tRNA. These trigger inflammation through their telomeres.

G U CC U G
C U GG U C

Although these have a Y chromosome, they do not have an X chromosome.

The telomeres are responsible for the rod shape of bacteria when the bacteriophages attach themselves to spores and provide the spore with their cytoplasm through their telomere as soon as the bacteriophages have pierced the spore or cell with their telomere through the cell membrane. The telomeres are responsible for the rod shape of bacteria when the bacteriophages attach themselves to spores and provide the spore with their cytoplasm through their telomere as soon as the bacteriophages have pierced the spore or cell with their telomere through the cell membrane. I have already seen a video on YouTube from the channel: "IrgendwasMitArte" or "Irgendwas Mit Arte."

I am right in saying that guanine has made up NH_3CO_2. Because from this element emerges the hydrocarbon.

The so-called macrophages are responsible for the spore stage of living beings so that bacteriophages can bring about a corresponding transformation or mutation. Macrophages are coronaviruses because their peplomers are a prerequisite for the spore stage. All coronaviruses are retroviruses and can multiply outside of hosts, e.g., fungi, etc. That is also why caterpillars go into the spore state, so to speak, into the shape of dormancy to transform.

T U CC U TT U C
C U TT U CC U T

Here, too, only Y chromosomes are present, but no X chromosomes.

Anyone who wants to assume that I have no idea, despite or precisely because of the many repetitions, errors, confusions, and contradictions should try to find out whether this is written somewhere in a textbook and should think about how far geneticists and virologists were able to achieve notable successes with their previous knowledge. Namely, none, and therefore many Bullshits has spread, and only placebo products have been offering as vaccines or medicines.

A genome has been creating through tRNA and mRNA interaction, i.e., between bacteriophage and macrophage. However, the critical point is that the result, which type of living being or plant emerges from this interaction, can be attributed to pure coincidence, which repeatedly leads to so-called errors in reproduction.

I would also like to note that a gram-negative does not produce a green color since phosphorus is always positively charged. The green color stands for the so-called positron, which means that when two protons collide and an electron beam hits them simultaneously, you get a positively charged electron, better known as a positron.

Ultimately, the way of life and behavior of all people, concerning their environment, are decisive, whether pandemics, epidemics, etc., arise. Anyone who surrenders to nature or the filth in their urban living environment will have to draw the consequences for their actions accordingly. If it is in the form of a fatal disease, then it is a deadly disease.

At this point, I would like to provide an example of the relationship between an RNA and a DNA and the reason, perhaps also proof, why an mRNA and tRNA depend on the respective RNA of the Y chromosome and X chromosome.

The condition of the first example is the RNA of the Y chromosome, and therefore I only use one element as tRNA to replicate the RNA of the Y chromosome to show what could happen:

<u>RNA of the Y chromosome:</u>

G AT C

Now the arbitrarily chosen tRNA, which only has one amino acid of a tRNA:

C

The following results could occur, whereby both results represent a genetic disposition:

G AT C
C CC C

Or

G AT C
C ■■ C

It should have been noting here that the two black squares have intended to indicate that no conversion has taken place or could occur in this area so that a gap occurs here in the copied second

RNA strand. The first result seems far more improbable than the second but cannot be ruled out without experiment.

Now a second example so that you can better understand what I mean:

This time I take the RNA of the X chromosome and use the mRNA, two amino acids of the RNA of the X chromosome, whereby the principle applies to both male and female DNA. So that a tRNA can also achieve the result that arises concerning the Y chromosome.

RNA of the X chromosome:

G AC U

Now I am using two amino acids from the X chromosome RNA as mRNA:

A U

We also get a genetic disposition for the subsequent two cases:

G AC U
C UU C

Or

G AC U
C ■U C

Conclusion: To produce a tRNA and mRNA, you need at least two amino acids of the RNA of the Y chromosome and X chromosome to be able to subject two nucleotides to a complete reverse transcriptase because you can see very nicely how the amino acids C and U become one again X chromosome, which would not be the case if only one amino acid has used.

I recently found out that the so-called coronaviruses are nothing more than host cells, i.e., retroviruses, and why these can only replicate by presenting their RNA.

ORIGIN OF LIFE IN ITS ORIGINAL FORM

1) Viruses are the basis of the cell membrane or form the basis on which the RNA and DNA structure has ultimately been building.

2) Viruses trigger catalysis by absorbing hydrocarbons.

3) When Einsteinium emits helium, it creates hydrocarbons. Einsteinium then forms the atomic shell and helium the atomic nucleus.

4) When helium emits hydrocarbon, helium is completely dissolved, leaving hydrocarbon.

5) Affectations arise from a lack of catalysis, or affectations are favored by it, which leads to a lack of insight from a psychological point of view.

6) When hydrocarbons correlate (react) with phosphorus, photosynthesis is triggered. But so long as there are no proteins, phosphorylation cannot take place.

7) From the condition that leads to photosynthesis, only the DNA (German: DNS) results (corresponds), whereby the DNA (German: DNS) itself has no structure, but only exists because radium is formed as an absorption product during photosynthesis, without that no potential energy would exist.

8) Without photosynthesis, humans change their being to evil and become a psychopath from a psychological perspective.

<u>Annotation:</u>

Am I always right with my theories or not? I will leave that to "legends."

For example, to obtain guanine, you need the juice of a cucumber. This juice is then heated, and the steam is then cooled and captured. For Obtaining thymine, the tomato has been using in the same way, a lemon for uracil, a plum for adenine, and an orange or tangerine for cytosine. The juice must have heated because the pH value of the juices from fruits and vegetables is too high to obtain a suitable vaccine from it. The heating will cause all the carbon to disappear, which will lower the pH value. As soon as the pH value is at zero, you can switch over and make a serum for a vaccine as with water. The serum only serves as a nutrient medium. A virus or bacterium must have been introduced into this serum to have an RNA or DNA template available. As a comparison, I would then take a sip of the serum and then compare the taste with the current vaccines to see whether it tastes the same. I do not know whether there will be a difference in the color component after heating the fruit and vegetable juices. Still, there are not only objective but also subjective properties that can have used for comparison. I almost forgot that the reason for a neutral pH value is that no thrombosis occurs when serum or vaccine has been administering. As the pH value increases, so does the number of thromboses, suggesting clumping the blood, telling that the acid content is too high.

The difference between fruits and vegetables in their respective amino acid bases is that vegetables provide carbohydrates since the amino acid bases are carbon-based when the vegetables ripen in the soil. Carbohydrates are also known as glucose. On the other hand, Fruit provides proteins because the amino acid bases have based on hydrogen, which is then present as glycose, provided that the ripening process takes place on the surface and not in the soil.

No rapid antigen test or PCR test method has ever produced a positive result. The results have deliberately misinterpreted - I claim out of ignorance, which virologists are embarrassed to admit and do not want to gamble away the chance to distinguish themselves. As already mentioned, the hospitals in China were understaffed, but this has suddenly changed due to the pandemic. And I know that for sure, even though I have never been to China.

Even if I repeat myself, I am now sure of the difference between RNA and DNA viruses. RNA viruses are nothing more than bacteria and are therefore rod-shaped. In contrast, DNA viruses are host cells and are consequently circular. Host cells also referred to as retroviruses, can reproduce outside of another organism due to their double helix genome and only need a suitable nutrient medium. Water is so essential as the basis of all life. However, bacteria cannot multiply outside an organism and need a host cell (DNA virus). Therefore, everything else I have written so far, I can confidently discard.

What annoys me the most is that Mai Thi Nguyen-Kim Leiendecker is hiding her knowledge of the development of hydrogel-based mRNA vaccines and keeping a great secret about it concerning her Ph.D. Anyone who deals with your doctoral thesis on the subject of "hydrogels" and their applications will find that your doctoral dissertation is excellent and offers entirely new perspectives for all areas of life - without having to highlight a specific location. She explained the development of an mRNA vaccine very well and - also in the illustration - because you only need to take stem cells from sick patients or non-sick patients and bring them together with a virus so that the virus through the corresponding hydrogel has prevented from preventing cell division from being able to survive itself so that after a fusion of stem cells and pathogens, these are administered to the corresponding patient together. I do not want to go into further details now because the work is pervasive. Of course, I have continued your thoughts regarding the development of mRNA vaccines - although the application of the hydrogels is exceptionally versatile. If you have a lot of money and do not know what to do with it, you should devote yourself to this doctoral thesis and try something for the world and your fortune. I should change the title, as I now know that a DNA virus uses the peplomers, which make up the coronavirus ring, to attach itself to other host cells. Even if the coronavirus is a host cell, it differs from which organism the host cell originates. An animal host cell will attach itself to a human host cell to deprive the human host cell of the nutrients that an animal host cell needs to multiply by division, which prevents the human host cell from dividing itself. That is what has meant by overwriting DNA. DNA has not been overwriting, but a host cell that is foreign to the body removes the nutrients from the body's host cell to begin a corresponding transformation.

I believe that I now understand the difference between traditional vaccine and mRNA vaccine, which I would never have achieved without Mai Thi Nguyen-Kim Leiendecker. In the conventional vaccine, the virus or bacterium that causes a particular disease in humans has been administering. Still, for an mRNA vaccine, stem cells must have been taking from each infected. These must then be placed in a culture medium, which is the DNA of the virus or RNA of the bacterium and are cultivated in this nutrient medium so that the body's stem cells are then genetically modified and given back to the owner since a virus or bacterium would never eat each other because with an mRNA vaccine the pathogen (virus or bacterium) can be fooled into believing that it is a conspecific. Also, I think I have understood what the purpose of telomeres and peplomers is. Telomeres require a bacterium or an insect, a plant or human to defend themselves against predators, peplomers, on the other hand, require a virus or insects, plants, and humans to absorb nutrients, but also to excrete substances or to be able to mate and multiply via the peplomers.

Even though I repeat myself personally, I am now 100% sure that coronaviruses are not pathogens but only stem cells. The reason is that the chemical composition of the "coronaviruses" is identical to that of the chemical composition of the stem cells. However, for a virus and bacterium to pose a danger, viruses and bacteria would have to consist of entirely different chemical substances that do not resemble the stem cells of humans, animals, and plants at all. June Almeida has not discovered a virus but stem cells.

I got my first vaccination today, and I do not feel anything. I did not even feel the injection. The whole theater about the vaccines has no basis. If the vaccines led to said thrombosis, then the arm in which the injection has been administering would have to swell, which will never be the case and never will be the case. No matter which vaccine you have been giving, none of the vaccines will cause any side effects. All the studies around possible side effects are obsolete because no study has ever yielded adverse results but is only pure imagination. Instead, I do not think any studies have been carried out but have only taken place on paper. You can safely vaccinate anyone with any vaccine, whether expectant mothers or people with pre-existing conditions, even adolescents and infants. And there will also be no interactions with medication if the one vaccinating due to a chronic disease must take medication. Many scientists, physicians, journalists, etc., should have a psychiatric report prepared so that they know whether they are socially acceptable at all or not.

The whole pandemic is pure money-making and nothing else, with all the vaccines just placebo products. Now I also know why the placebo effect can very well cause a cure. The placebo effect has been using to trigger ATP synthase through lysis. But what many citizens underestimate are the influenza A viruses, which can lead to coronary heart disease in every person. Therefore, the dangerousness of coronaviruses - regardless of whether they are stem cells or not - cannot be ruled out. When virologists or medics like Dr. Bhakdi - as in his book: "Spectre Infections" (only available in German) - trigger the smallpox virus shingles, it shows that virologists again have not studied well either. I learned that poxviruses and shingles start smallpox by herpes viruses. Even authors like Walter van Rossum should stop spreading any conspiracy theories if they have no idea about infections and the like. My book title also reveals a conspiracy theory background, but I preferred to deal with the scientific experience instead of politics. In addition, I will prove that even people like Christian Drosten, Hendrik Streeck, and others have no idea about mathematics. Suppose a nanometer has a millionth of a millimeter and a coronavirus size between 60 and 120 nanometers. In that case, this virus cannot be detected with a conventional microscope but can only be detected using transmission electron microscopy.

And for this reason, Christian Drosten cannot have discovered the SARS virus at all, as he probably missed this magnitude. Therefore, anyone who can drink, smoke, take drugs, mess with children, or create such things does not need to worry about the side effects of vaccinations, as people do not worry about their perverse machinations. And as already mentioned, I should not take viral infections lightly, as I also got shingles in midsummer in 2013, which disfigured my face and caused my brain to swell. But thanks to the antibiotics, the infection was over after ten days, but the vacation was spoiled. What also needs to have been considering is that an ATP synthase through lysis is a prerequisite for phagocytosis. Phagocytosis is the cell metabolism that has necessary to cleanse the organism of all pollutants and toxins. The term polio is the medical name for polio and has been triggering by the rubella virus and personal misconduct during pregnancy.

I also learned that adhesive forces are at work to create DNA but to separate a strand into two RNA strands, cohesive troops have required. All virologists in the world do not want to admit that diseases are due to a lack of nutrients, pollution, and overpopulation, which leads to a lack of food. Diphtheria, for example, is only given to those who drink dirty water. Diphtheria did not go away with any vaccinations, but solely because people in the western world have clean drinking water. Tetanus only occurs when amputations perform with dirty cutlery, which is no longer the case these days. Much of what we attribute to viruses and bacteria is our own through our life and behavior. The cause of diseases among people is always the person himself - no matter how much people talk about viruses and bacteria.

A vaccine does not improve the living conditions in which most people live. Regardless of what I will tinker with, my point of view about viruses and bacteria, only my way of life and behavior are the basis of the quality of the living conditions. And that is what nobody in their twisted western head wants to see and accept. Nobody on earth needs drugs or vaccines if the appropriate lifestyles and behaviors contribute to maintaining health. But most people - including me - display lifestyles and behaviors that are much more conducive to self-destruction than self-preservation. Regardless of how many books have written about viruses and bacteria, vaccines, and drugs, we will have to change our way of life and behavior and curb the multiplication. Because the more people live on this planet, the faster resources have used up, which leads to unrest and conflict, and neither a vaccine nor a drug, books, or anything else helps. Also, as the world population increases, pollution is more likely to increase than decrease. Because the more people, the more difficult it becomes to get people to understand. But no pharmaceutical company will make money if it tries to persuade people to improve their life and behavior. Not even doctors or medical professionals are interested because they would have to fear for their jobs like in China if doctors and medical professionals in hospitals had nothing or truly little to do before the pandemic. Who knows how much of what is in the books and libraries of this world on the subject: Viruses and bacteria have based on objective points of view? I claim that most of it is just pure speculation on the part of virologists and biologists, etc., and accordingly, I let myself, of course, be tempted to take some Bullshit at face value. Most of it is based only on subjectivity, that is, on interpretations, but not on observations.

I could throw everything that has been written so far in the bin because even the virologists and epidemiologists contradict each other, which suggests that they all have no idea about the essential things in life and would much rather be in the spotlight to find out more about themselves to live like stars on the red carpet. But I leave everything as it is to show how views and insights change over time. Because for me, I can say that the likelihood of dying from lung cancer from smoking is more significant than from any suggested side effects of any vaccine. When virologists and epidemiologists like Prof. Dr. Hendrik Streeck claim that the loss of the sense of taste and smell is due to a COVID-19 illness, they probably forgot that all influenza A viruses cause the same flu symptoms. No general practitioner can tell the multitude of flu illnesses apart, as there are no so-called rapid tests for most influenza A viruses, but now of all times for COVID-19. In addition, due to the lysis, an ATP synthase is responsible for the immune system and emotions. The ATP synthase through lysis is required to convert one strand of DNA into two strands of RNA to correct the genetic material with the help of adenosine dehydrogenase. The ATP synthase from lysis is when you break a sweat, causing the carbon bonds to dissolve to allow cells to divide.

Interestingly, Wikipedia says so much about stem cells, viruses, etc., that I can hardly believe it has not been considering. It is more likely that there is no real objective evidence of what is in it. To trust Wikipedia completely, one should be careful because the pages have only been creating by laypeople who publish everything on a topic without scientific review, just because it is in a book somewhere. If everything in Wikipedia were truthful, then all libraries could close. Antigens only enable lifelong immune protection when an adenosine dehydrogenase forms the ATP synthase.

In contrast, forming antibodies by an ATP synthase through the lysis only offers temporary protection, which means renewing the protection regularly. At what time interval, I do not know. Cancer cells have been defeating by taking stem cells from the patient, cultivating them in a hydrogel, and injecting them directly into the cancer cells. What still needs to be mentioning is that stem cells that have only been producing in a nutrient solution offer insufficient protection against renewed infection or are unsuitable for cancer treatment. In contrast to stem cells cultivated and administered in a hydrogel, they offer lifelong protection against cancer cell formation. Of course, the patient must show a certain degree of personal responsibility and change his lifestyle and behavior accordingly. With the help of the cultivation of stem cells in hydrogels, one can generally restore defective organs so that a transplant would be superfluous. Some time ago, I downloaded the instructions with the title: "MANUEL FOR RAPID LABORATORY VIRAL DIAGNOSIS" from the German language Wikipedia, the WORLD HEALTH ORGANIZATION GENEVA published in 1979. The second chapter of this manual describes dealing with viruses that cause respiratory diseases and how they can be detected directly and indirectly. These are seen now with the aid of electron microscopy, indirectly with immunofluorescence. It describes precisely the scenario that we are experiencing today. The following people developed these instructions: 1) June D. Almeida (from The Wellcome Research Laboratories Beckenham, England), 2) D.W. Bradley (from Hepatitis Laboratories Division CDC Bureau of Epidemiology Phoenix, AZ, USA), 3) J.E. Maynard (from Hepatitis Laboratories Division CDC Bureau of Epidemiology Phoenix, AZ, USA), 4) A. Voller (from Nuffield Laboratories of Comparative Medicine Institute of Zoology London, England), 5) P. Atanasiu (from Institute Pasteur Paris, France), 6) P.S. Gardner (from Department of Virology Royal Victoria Infirmary Newcastle upon Tyne, England), 7) A.W. Schuurs (from Organon Scientific Development Group Oss, Netherlands), 8) R.H. Yolken (from Johns Hopkins University Medical School Department of Pediatrics Baltimore, MD, USA). Since I now know that my theories cannot have been refuting, I can confidently go on. If virologists, epidemiologists, science journalists, conspiracy theorists, politicians, etc., cannot even do proper research, I can also book my theories. I cannot go wrong if scientists do not do their homework or are not up to their task despite the immense libraries. I am also confident that genetics is still in its infancy, and Americans know nothing about genetics. They speculate, otherwise, there would be no physical or mental disabilities in America, nor any other types of diseases, and Americans would then be able to live an average age many times higher than the rest of the world and would also be the first to put a vaccine on the market. Still, they slept because they did not know.

Respiratory diseases must also include allergies such as hay fever, which leads to inflammation of the nasal mucous membranes and throat, and asthma since asthmatics must have chronic pneumonia anyway. But shortness of breath can also be a sign of coronary heart disease. An ATP synthase by an adenosine dehydrogenase has also formed because macrophages eat bacteria to use their material to produce adenine and produce new macrophages. The macrophages release adenosine. Adenosine has made up of adenine and cytosine, from which the name adenosine has derived.

On the other hand, the lysis forms an ATP synthase because bacteriophages destroy viruses by positioning themselves on the spores to inject guanine into the viruses then. The destruction of the spores by the guanine releases the so-called guanosine. Thus, guanosine has made up guanine and cytosine, which is also why the derived name guanosine. As a result, immunofluorescence has been producing by both guanosine and adenosine. Thus, the body controls the blood flow via immunofluorescence. It should also have been noting here that adenine is required, for example, to produce arteries. It is also one of the reasons why adenine gets the blue color associated with the arteries. Therefore, I would recommend an active vaccination as a first vaccination and a passive vaccination as a second vaccination. The primary vaccination has based on the principle of the formation of antibodies when catabolism occurs. For this, guanosine has needed to cultivate the virus in it. The second vaccination has based on the principle of the formation of antigens when anabolism occurs. In this case, adenosine has required, and the virus has cultivated in it. However, a single vaccination would also be sufficient by mixing adenine + guanine + cytosine to produce the adenoguanosine derived from the term adenosine (dehydrogenase) - in which the virus has then cultivated. Adenoguanosine is needed to naturally develop steroids so that the body and mind can perform better. Coronaviruses, also known as retroviruses or geneticists, refer to them as stem cells and can be of vegetable, animal, or human origin. Coronaviruses (retroviruses, also known as stem cells) only make people sick if a non-species-specific coronavirus, i.e., retrovirus or stem cell, passes from an animal to a person or vice versa. The same also applies to plant coronaviruses (retroviruses or stem cells) such as fungi, etc., which cannot have been using by the body to form protein synthesis, since their stem cells also contain inorganic elements that are not edible for humans, for example, are for instance toadstool.

In contrast, there are also edible mushrooms for humans and animals. But some animals are inedible for humans, such as pufferfish or worms or bats, etc., but on the other hand, some animals are edible. Rossana Segreto and Yuri Deigin claim in their essay: "The genetic structure of SARS-CoV-2 does not rule out a laboratory origin" those chimeric viruses arise through natural recombination or human intervention. All types of coronaviruses are chimeric viruses. The term chimeric stands for hermaphroditic since all coronaviruses are retroviruses anyway and only multiply by themselves. Coronaviruses can be of human, animal, and vegetable origin. A chimeric (hermaphrodite) property is the prerequisite for developing chirality (mirroring of the limbs). In addition, natural recombination has also known as reverse transcriptase.

At this moment, a new idea of the possibility of interpreting genetic structures occurs to me again:

G A GA T CT A G
A G AG A TC T A

<u>The following Y chromosomes now result:</u>

G A GA T T A G
A G AG A T T A

1) <u>two Y chromosomes twice:</u>

It must have been reading as follows: G and A must be read twice "fictitiously" in one case, which means a value has been reading twice.

<u>first Y chromosome:</u>

<u>second Y chromosome:</u>

The blue squares are synonymous with A - for reasons of clarity. I also crossed out the G that has been reading twice to know which G it is.

<u>third Y chromosome:</u>

<u>the fourth Y chromosome, with the double read A, also have been crossing out:</u>

The fourth Y chromosome even has the property that it even has A crossed out twice.

<u>fifth Y chromosome:</u>

The fifth Y chromosome also has the property of two A's that have crossed out.

<u>We even get two additional Y chromosomes from G and A:</u>

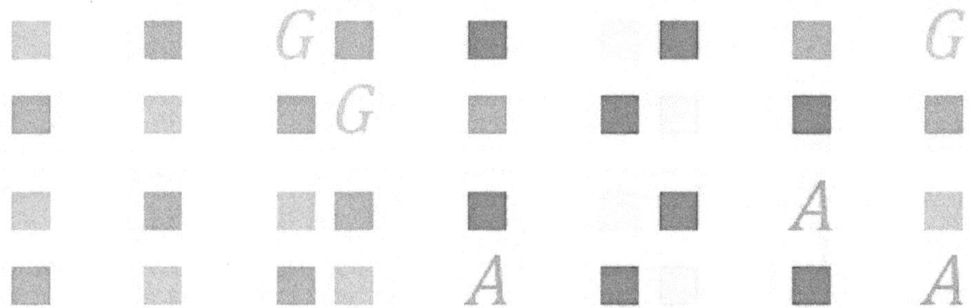

<u>ninth and tenth Y chromosomes, which are also nested with one another:</u>

T *T*
T

<u>tenth Y chromosome:</u>

T
T *T*

Now we move on to the X chromosomes, where there are four of them in total, three of which are nested with one another:

<u>first X chromosome:</u>

G *A*
A *G*

<u>second X chromosome:</u>

A G
G A

third X chromosome:

G A
A G

fourth X chromosome:

C T
T C

A reverse transcriptase already occurs in the first row and not, as with me until now, only in the second row. So, combining all the same elements always gets a harmonic oscillation of at least two rows.

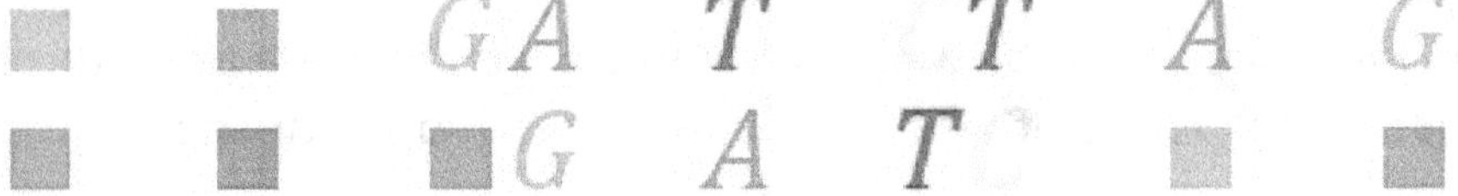

We get G-A-T-C and C-T-A-G in the first row and G-A-T-C in the second. The principle of genetics is like the Rubik's Cube. Today, I realized that genetic sequences have no polarity but that the contradiction of genetic sequences means the respective reading direction. Positivity of a genetic sequence implies nothing more than reading the genetic structure from right to left, from the outside to the inside. The reflection that I always get through a reverse transcriptase so that a chirality becomes possible is because the negativity takes place in a changed reading direction from left to right. It would help if you imagined it that way. The positive picture is always the accurate picture. The negative image is always the mirror image. Namely, the positive image, i.e., the actual image, is on the right, and the negative image, i.e., the mirror image, on the left.

Coronaviruses cannot have crossed because coronaviruses do not combine but always only divide to produce an organism and the external conditions than the rest. I had probably misunderstood the illusion, but in the end, it did not matter whether the chimera is a cross between two types of viruses or not, as viruses do not combine or with foreign viruses. Because of this, the assumption that the SARS-CoV-2 virus is a cross, i.e., a delusion, is just metaphysical nonsense. A pathogen in animals can only exist as soon as the animals themselves suffer from a disease. After that, the pathogen can have to transmit from one animal to another or a person. It is also the case with humans and plants. Suppose neither a person nor a plant is infected. In that case, neither a person, plant, or animal poses any danger - except insects, which do not transmit viruses, but only Poisons that can cause inflammation in humans and animals. Virus infections only ever take place within a body, as viruses are hydrophilic. Bacterial infections, on the other hand, always take place outside the body because bacteria are aerobic.

On the other hand, Fungi spores are hydrophobic, as they are satisfied with a damp environment (e.g., fungal infestation after water ingress in the apartment) and a dry environment. Whether hydrophilic, aerobic, or hydrophobic properties have developed constantly depends on the respective living conditions. When I wrote the book, to create a better connection to the pandemic events, I had the title without a question mark in my head, but meanwhile, it has become a question mark. I would also like to explain why I have discovered that, e.g., chickenpox has caused coronaviruses. It can have been recognizing by the fact that a white wreath appears around a pimple. This white wreath arises when a coronavirus is at work, whereby the wreath becomes white because nitrogen is forced outwards by the peplomers. For me, however, it is also an indication that coronaviruses attach themselves to a cell through the peplomers to either supply it with nutrients or to remove nutrients from the cell. Avian flu has based on a misunderstanding. The misunderstanding is that the German term for "Windpocken" is called "chickenpox" in English. That is also why I turned away from my initial mistrust, which I can only do because I dealt with the topic "Viruses" myself. It is irrelevant to what extent I make the right connections since I am just a layperson and not an expert. The coronaviruses discovered or their mutations do not necessarily have to come from bats. It is enough if human coronaviruses, i.e., retroviruses or stem cells, are defective, e.g., due to an inflammation in the own body, leading to a disease in a healthy person if these are transmitted. It is therefore essential that people go to the dentist regularly, for example, because stem cells (coronaviruses, i.e., retroviruses) are in the teeth, as these have also been using for genetic analyzes, and if the teeth are rotten or teeth are infected. It can contagions occur, leading to illnesses in others, as the corresponding pathogens then make their way through droplets via the nasal and oral mucous membranes into the respiratory tract. The same applies to people with colds, chronic bronchial asthma, etc.

I have looked at the subject of "bird flu" again retrospectively and noticed that the birds themselves fell victim first; died. It means that the SARS virus was and, of course, still dangerous for humans as a result. At this point, I also claim that if the SARS-CoV-2 virus originates from bats, difficulties must arise in the animals themselves to lead to considerable health problems in humans. But it's partly your fault if you eat any shit.

At this point, I would like to introduce another extension. The fact is that an ATP synthase produces antibodies through the lysis, but the effect is only short-lived, max. 72 hours. That is also why the vaccine should only ever have given to patients whose disease has already broken out. Since the vaccine triggers an ATP synthase through lysis, it must be administered every three days until curing the disease. The principle of ATP synthase through lysis, which creates lysosomes, is confirmed by my experience in the field of sports, as the guide has based on the following: the first day of heavy equipment training, the second day of cardio workout, and the third day of break and then everything again front. I am slowly getting fed up with discoveries. The lysosomes have been using to decipher the RNA sequence and the ribosomes to decipher the DNA sequence. Ribosomes create the so-called antigens and lysosomes, the antibodies - even if I repeat myself. The last insight I would like to put on paper, for the time being, is that antibodies are responsible for renewing the cytoplasm, and antigens are responsible for renewing the cell membrane. The reason female DNA evolved from tRNA is that the term tRNA stands for thymine-based RNA. In the beginning, the egg cell consists only of thymine, so that cells build or have a cell membrane at all. The male DNA, i.e., mRNA, on the other hand, supplies the protein-coding enzyme to convert the thymine into cytosine.

I think I may now know why the rapid antigen tests do not work for laypeople because 1) there is no centrifuge available to increase the vesicle proportion to obtain detection by ELISA even without a PCR method. And 2) because a proper primer did not use. The reason for this assumption is that on 07/19/21, I read the following article in the journal: "Spectrum of Science - The Secrets of the Immune Defense" (only available in German), which I would like to use in the original, even if I might get in trouble. The article is about "Long-lived memory T-cells." It reads as follows: “Human subjects received a yellow fever vaccine, which produces long-lasting immunity, and drank a predetermined amount of heavy water for the following 14 days. The blue-colored area in the diagram shows an increasing accumulation of deuterium markings in newly formed, yellow fever-specific cytotoxic T cells obtained from the test subjects' blood during the first 28 days after the vaccination. The measurement curves have shown record the survival of the marked T lymphocytes in different patients. As the cells die, the accumulation of the markings decreases again; the mean half-life of this process was about 460 days. In addition, the rate of division of the marked cells was less than one per year. These results show: In the first two weeks after the vaccination, there was an intensive increase in T lymphocytes, which resulted in yellow fever-specific cells that have proven to be remarkably long-lived." To be able to use a rapid antigen test correctly, a proper primer is required first. I would try the following primer, and it should consist of deuterium and a ribose 5'-phosphate. After a swab has been taking from a subject, the sample has centrifuged. It is then filtered and can have been using for the PCR method. I cannot say how long a model must have been centrifuging to obtain enough DNA material. The ribose 5'-phosphate, which can also confidently be called testosterone, serves as a protein-coding enzyme. In doing so, I found another article about the protein-coding enzyme in the journal of the same name, which has the title: "On the trail of immune memory" and reads as follows:" When cells divide, the DNA in the nucleus doubles; the newly formed genetic material is distributed among the daughter cells (T lymphocytes and have a thymine-based RNA) and has retained over their entire lifespan. Suppose you add deuterium-containing glucose or heavy water during the division (red in the picture). In that case, the atoms of this isotope are incorporated into the newly emerging DNA strand and label with it. The markings are lost again when the daughter cells die, and their genetic material disintegrates. However, if the cells divide later and their genetic material has distributed to the next generation, the marked DNA building blocks have diluted accordingly. By determining how the amount of DNA markers in a cell population evolves, one can find out something about the lifespan of the cells." The entire article has based on the following sources:

1) Barouch, D.H. et al.: Correlates of protection against SARS-CoV-2 in rhesus macaques. Nature 590, 2020

2) Hellerstein, M. et al.: Origin and differentiation of human memory CDB T cells after vaccination. Nature 552, 2017

3) Hellerstein, M. et al.: Measurement of cell proliferation by heavy water labeling. Nature Protocols 2, 2007

Furthermore, damage to the X and Y chromosomes can have been repairing, whereby T helper cells have required this. For repairing cracks to the Y chromosomes, a nutrient medium has needed, consisting of glucose-6-phosphate, in which the individual stem cells of the patient are cultivated and then administered. The glucose-6-phosphate is also necessary to find out whether the Y chromosome is damaged at all. On the other hand, to eliminate X chromosomes, a culture medium consisting of ribose-5-phosphate is required to cultivate and administer the individual stem cells of the affected person. Here, too, ribose-5-phosphate is used to identify damage to the X chromosomes. The following should also be noted here: 1) Glucose is composed of guanine, uracil, and cytosine. 2) Glycose, on the other hand, is made up of guanine, thymine, and cytosine. 3) Glycogen is made up of guanine, thymine, and adenine. In addition, the T helper cells cultivating in a glucose-6-phosphate culture medium must display T in a rapid antigen test, and C must have been demonstrating in the case of the ribose-5-phosphate medium. I am talking about the Y chromosome of tRNA, thymine-based RNA, because the heart has based on it.

I claim that I have now found out how a division and thus inheritance takes place, and I would like to anchor this as a law:

Law on the division encryption of DNA and RNA of any kind

A Tribute To Dr. Melanie Brinkmann

and Zhou et al.
(from the Wuhan Institute of Virology (WIV))

It should have been anticipating that a division occurs at that point where a value appears twice in a row.

$$\begin{bmatrix} G & \rightarrow & A \\ \uparrow & \ddots & \vdots \\ A & \cdots & T \end{bmatrix}\begin{bmatrix} T & \rightarrow & C \\ \vdots & \ddots & \downarrow \\ C & \cdots & T \end{bmatrix}\begin{bmatrix} C & \rightarrow & T \\ \uparrow & \ddots & \vdots \\ T & \cdots & C \end{bmatrix}\begin{bmatrix} A & \rightarrow & G \\ \vdots & \ddots & \downarrow \\ T & \cdots & A \end{bmatrix}$$

$$\begin{bmatrix} T & \cdots & C \\ \uparrow & \ddots & \vdots \\ C & \leftarrow & T \end{bmatrix}\begin{bmatrix} T & \cdots & A \\ \vdots & \ddots & \downarrow \\ A & \leftarrow & G \end{bmatrix}\begin{bmatrix} A & \cdots & T \\ \uparrow & \ddots & \vdots \\ G & \leftarrow & A \end{bmatrix}\begin{bmatrix} C & \cdots & T \\ \vdots & \ddots & \downarrow \\ T & \leftarrow & C \end{bmatrix}$$

We get the following Y chromosomes, whereby these always together form a Merkel diamond, and for the sake of simplicity, I switch to a different matrix view:

<u>1) The first two Merkel diamonds are as follows:</u>

Two Merkel diamonds always mean four Y chromosomes.

$$\begin{bmatrix} & \cdots & \\ \vdots & \ddots & \vdots \\ & \cdots & T \end{bmatrix}\begin{bmatrix} T & \cdots & \\ \vdots & \ddots & \vdots \\ & \cdots & T \end{bmatrix}\begin{bmatrix} & \cdots & T \\ \vdots & \ddots & \vdots \\ T & \cdots & \end{bmatrix}\begin{bmatrix} & \cdots & \\ \vdots & \ddots & \vdots \\ T & \cdots & \end{bmatrix}$$

$$\begin{bmatrix} & \cdots & \\ \vdots & \ddots & \vdots \\ & \cdots & \end{bmatrix}\begin{bmatrix} T & \cdots & \\ \vdots & \ddots & \vdots \\ & \cdots & \end{bmatrix}\begin{bmatrix} & \cdots & T \\ \vdots & \ddots & \vdots \\ & \cdots & \end{bmatrix}\begin{bmatrix} & \cdots & \\ \vdots & \ddots & \vdots \\ & \cdots & \end{bmatrix}$$

<u>2) The second two Merkel diamonds are as follows:</u>

Here, too, two Merkel diamonds mean four Y chromosomes.

$$\begin{bmatrix} & \cdots & \\ \vdots & \ddots & \vdots \\ & \cdots & T \end{bmatrix}\begin{bmatrix} & \cdots & \\ \vdots & \ddots & \vdots \\ & \cdots & \end{bmatrix}\begin{bmatrix} & \cdots & \\ \vdots & \ddots & \vdots \\ & \cdots & \end{bmatrix}\begin{bmatrix} & \cdots & \\ \vdots & \ddots & \vdots \\ T & \cdots & \end{bmatrix}$$

$$\begin{bmatrix} T & \cdots & \\ \vdots & \ddots & \vdots \\ & \cdots & T \end{bmatrix}\begin{bmatrix} T & \cdots & \\ \vdots & \ddots & \vdots \\ & \cdots & \end{bmatrix}\begin{bmatrix} & \cdots & T \\ \vdots & \ddots & \vdots \\ & \cdots & \end{bmatrix}\begin{bmatrix} & \cdots & T \\ \vdots & \ddots & \vdots \\ T & \cdots & \end{bmatrix}$$

<u>3) Now we come to the X chromosomes:</u>

$$\begin{bmatrix} \cdot & \cdot & \cdot \\ \cdot & \cdot & \cdot \\ \cdot & \cdot & \cdot \end{bmatrix}\begin{bmatrix} T & \rightarrow & \\ \uparrow & \swarrow & \downarrow \\ & \leftarrow & T \end{bmatrix}\begin{bmatrix} & \rightarrow & T \\ \uparrow & \searrow & \downarrow \\ T & \leftarrow & \end{bmatrix}\begin{bmatrix} \cdot & \cdot & \cdot \\ \cdot & \cdot & \cdot \\ \cdot & \cdot & \cdot \end{bmatrix}$$

$$\begin{bmatrix} T & \rightarrow & \\ \uparrow & \swarrow & \downarrow \\ & \leftarrow & T \end{bmatrix}\begin{bmatrix} \cdot & \cdot & \cdot \\ \cdot & \cdot & \cdot \\ \cdot & \cdot & \cdot \end{bmatrix}\begin{bmatrix} \cdot & \cdot & \cdot \\ \cdot & \cdot & \cdot \\ \cdot & \cdot & \cdot \end{bmatrix}\begin{bmatrix} & \rightarrow & T \\ \uparrow & \searrow & \downarrow \\ T & \leftarrow & \end{bmatrix}$$

Some time ago, I claimed that different coronaviruses (retroviruses, stem cells) do not combine, but I noticed that it is possible after looking at the picture. If other coronaviruses (retroviruses, stem cells) connect, a new coronavirus (retrovirus, stem cell) has created. So, Rossana Segreto and Yuri Deigin are right. In the picture below, you can see very well how the coronaviruses combine.

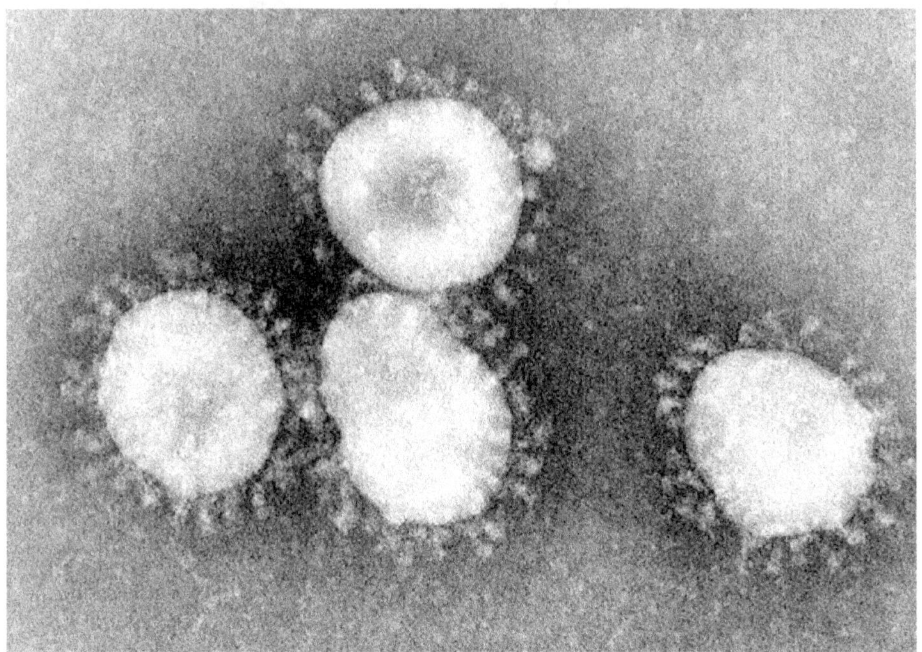

In the magazine: "Spectrum of Science - The Secrets of the Immune Defense," 8.21, p.19, article: "Smallpox - and a gift from cows," I read the following: "With a mortality rate of 30 to 50 percent, smallpox was once considered the world's deadliest infectious disease. The disease has caused viruses and has mainly been transmitted through the air by coughing or sneezing and body fluids, clothing, or bed linen. European settlers to the New World killed an estimated 30 percent of Native Americans, sometimes killing entire villages and tribes. Almost every settler town had its smallpox cemetery. Outbreaks have resulted in entire cities have been quarantining. Cows have been infecting. They can infect people with the cowpox virus, but this only causes relatively mild symptoms. Milking women who contracted cowpox develop bumps on their hands and arms. When the English country doctor Edward Jenner overheard a dairy farmer's story by chance, how glad she was to have such an infection That she did not have ugly smallpox scars on her face gave him an idea. In 1796 he developed the approach of injecting pus from cowpox pustules into healthy people to make these people resistant to the human variant of the disease. From today's point of view, this was a brutal and surprisingly unethical method. Still, it turned out to be effective: Contact with the cowpox material prevented the pathogen from becoming ill. In 1801 Jenner published his findings under the title: 'On the Origin of the Vaccine Inoculation.'" The term "vaccine" referred to the Latin word "Vacca" for a cow. The country doctor prophesied his method would ultimately lead to eradicating smallpox, the cruelest scourge of humankind. Jenner's discovery made him world-famous; politically, vaccinations were a charged issue from the start. Church people warned against interfering with the "divine creation" through vaccination. The crossing of the species barrier was criticized (as if humans had not consistently consumed animal products for millennia and surrounded themselves everywhere with animal products). Some politicians even called the vaccination "dirty witchcraft." Alone: She saved lives and in unimaginable numbers. An estimated 300 million people died of smallpox in the 20th century, but thanks to vaccination, there has not been a single known case worldwide since 1979. Jenner's prophecy had thus come true.

I first thought that measles could cause ugly facial scars with this article because I got it in school, but apparently, smallpox is to blame. So, you better take a closer look to see whether the smallpox is not itself coronavirus.

I also discovered different ideas for myself on page 20 in the journal: "Spektrum der Wissenschaft," which I take up to create several types of mRNA vaccines. With their help, you can develop therapies against cancer, like AIDS or dementia, or other physical and psychological ailments.

It has been noting here that I get a macrophage if I cultivate a virus in an appropriate culture medium. But you get a bacteriophage if you grow a bacterium in a nutrient medium. I come to this assumption because you shouldn't believe everything you read or what you see in documentaries since the representations of the bacteriophages look very far-fetched to me.

I would also like to start with a cell representation that shows what a cell of an inflamed tissue looks like, which, if left untreated, very quickly becomes a cancer cell. Such cells are called cytotoxic T cells (killer cells), which weaken the immune system. However, it has wrongly assumed that they protect the immune system.

Cytotoxic T cells can be viruses of all types that, instead of cytoplasm from cytosine, have cytoplasm from environmental toxins. Thus, if you were to cultivate a coronavirus in deuterium or ammonium nitrate, etc., and then inject them, you would quickly get evidence of when and why viruses and bacteria cause infections. In short: Viruses and bacteria cause disease because they absorb the corresponding toxins from the environment and then pass them on to the host. With this knowledge, of course, one also has a beautiful method of creating a medical-biological weapon.

Of course, a so-called cytotoxic T cell can be used to destroy cancer cells for medical purposes, but this must be introduced directly into the tumor. In this case, the cytotoxic T cells should prevent the destruction of healthy tissue during chemotherapy. In addition, they can have been reproducing particularly well and quickly by the X-rays. Electrochemical deoxidation - i.e., electrolysis - occurs in the cytotoxic T cells, commonly known as detox.

Law: Two arms of the immune system

A Tribute To Jason Becker

1) Humorous component:

a) **naive B cell:**

To produce a naive B-cell, one needs a virus or a bacterium to cultivate it in a culture medium of adenine + uracil + cytosine. As the viruses and bacteria multiply, the viruses and bacteria's properties from the nutrient medium have passed on to the next generations. Thus, each subsequent generation of viruses and bacteria has the corresponding properties.

b) **Memory B cell:**

Cultivating memory B cells, viruses or bacteria are required, developed in the following medium: guanine + adenine + uracil + cytosine.

c) **Plasma cell:**

It would help if you had viruses or bacteria cultivated in the following culture medium for this cell type: thymine + adenine + uracil + cytosine.

2) Cell-mediated component:

T helper cells:

a) **naive T-helper cell:**

Viruses or bacteria + culture medium from thymine + uracil + cytosine.

b) **Memory T helper cell:**

Viruses or bacteria + culture medium from thymine + guanine + uracil + cytosine.

cytotoxic T cells (killer cells):

a) **cytotoxic T cell (killer cell):**

Viruses or bacteria + culture medium from guanine + uracil + adenine + deuterium.

b) **cytotoxic memory T cell:**

Viruses or bacteria + culture medium from guanine + thymine + adenine + deuterium.

c) **Effector cell:**

Viruses or bacteria + culture medium from guanine + cytosine + deuterium.

The culture medium can, of course, have been modifying as desired. However, if coronaviruses have nothing to do with stem cells or if no stem cells emerge from coronaviruses, then you must take appropriate stem cells from animals or humans and let them into the relevant culture media to obtain proteins, which are then injected into the respective damaged organ so that the organ has repaired.

To come back to the individual amino acid bases. So, for example, guanine can have been making from glucose-5'-phosphate. This guanine from glucose-5'-phosphate can then have been using to obtain the other amino acid bases from it.

The following picture has intended to show what a cytotoxic cell looks like and distinguishes this coronavirus from others.

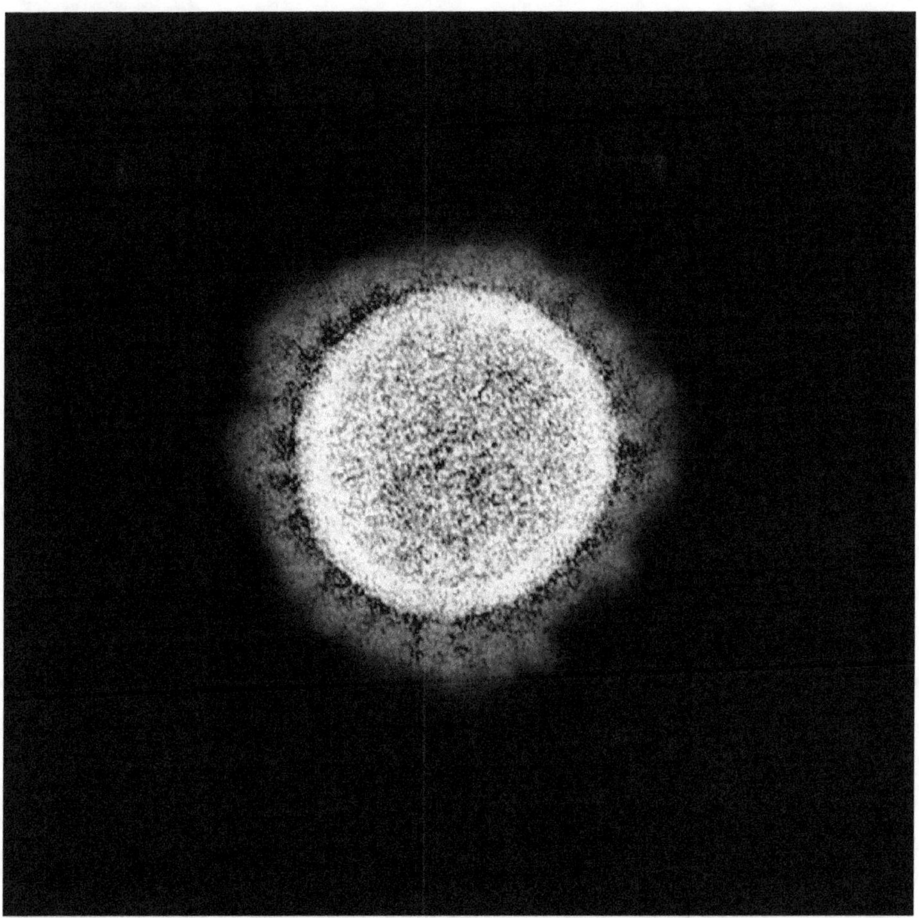

This coronavirus has neither thymine, nor adenine, nor guanine, nor cytosine, nor uracil.

I have once again discovered another method of analyzing RNA and DNA structures, which is as follows:

$$\begin{bmatrix} G & \rightarrow & A \\ \downarrow & \Searrow & \blacksquare \\ A & \blacksquare & G \end{bmatrix}\begin{bmatrix} T & \rightarrow & C \\ \blacksquare & \cdot & \uparrow \\ C & \blacksquare & T \end{bmatrix}\begin{bmatrix} C & \leftarrow & T \\ \uparrow & \cdot & \blacksquare \\ T & \blacksquare & C \end{bmatrix}\begin{bmatrix} A & \leftarrow & G \\ \blacksquare & \Swarrow & \downarrow \\ G & \blacksquare & A \end{bmatrix}$$

$$\begin{bmatrix} T & \blacksquare & C \\ \downarrow & \cdot & \blacksquare \\ C & \leftarrow & T \end{bmatrix}\begin{bmatrix} G & \blacksquare & A \\ \blacksquare & ⇘ & \uparrow \\ A & \leftarrow & G \end{bmatrix}\begin{bmatrix} A & \blacksquare & G \\ \uparrow & ⇙ & \blacksquare \\ G & \rightarrow & A \end{bmatrix}\begin{bmatrix} C & \blacksquare & T \\ \blacksquare & \cdot & \downarrow \\ T & \rightarrow & C \end{bmatrix}$$

$$\begin{bmatrix} C & \leftarrow & T \\ \uparrow & \cdot & \blacksquare \\ T & \blacksquare & C \end{bmatrix}\begin{bmatrix} A & \leftarrow & G \\ \blacksquare & ⇙ & \downarrow \\ G & \blacksquare & A \end{bmatrix}\begin{bmatrix} G & \rightarrow & A \\ \downarrow & ⇘ & \blacksquare \\ A & \blacksquare & G \end{bmatrix}\begin{bmatrix} T & \rightarrow & C \\ \blacksquare & \cdot & \uparrow \\ C & \blacksquare & T \end{bmatrix}$$

$$\begin{bmatrix} A & \blacksquare & G \\ \uparrow & ⇙ & \blacksquare \\ G & \rightarrow & A \end{bmatrix}\begin{bmatrix} C & \blacksquare & T \\ \blacksquare & \cdot & \downarrow \\ T & \rightarrow & C \end{bmatrix}\begin{bmatrix} T & \blacksquare & C \\ \downarrow & \cdot & \blacksquare \\ C & \leftarrow & T \end{bmatrix}\begin{bmatrix} G & \blacksquare & A \\ \blacksquare & ⇘ & \uparrow \\ A & \leftarrow & G \end{bmatrix}$$

To prove that you can't get very far in genetics without the German language, I want to provide an example for "AUG (E)":

$$\begin{bmatrix} A & \rightarrow & U \\ \downarrow & ⇘ & \blacksquare \\ U & \blacksquare & A \end{bmatrix}\begin{bmatrix} G & \blacksquare & A \\ \blacksquare & \cdot & \blacksquare \\ A & \blacksquare & G \end{bmatrix}\begin{bmatrix} U & \rightarrow & G \\ \blacksquare & \cdot & \uparrow \\ G & \leftarrow & U \end{bmatrix}\begin{bmatrix} U & \leftarrow & A \\ \blacksquare & ⇙ & \downarrow \\ A & \blacksquare & U \end{bmatrix}$$

$$\begin{bmatrix} G & \blacksquare & A \\ \blacksquare & \cdot & \blacksquare \\ A & \blacksquare & G \end{bmatrix}\begin{bmatrix} A & \blacksquare & U \\ \downarrow & ⇘ & \uparrow \\ U & \leftarrow & A \end{bmatrix}\begin{bmatrix} U & \blacksquare & A \\ \uparrow & ⇙ & \blacksquare \\ A & \rightarrow & U \end{bmatrix}\begin{bmatrix} U & \rightarrow & G \\ \blacksquare & \cdot & \uparrow \\ G & \blacksquare & U \end{bmatrix}$$

$$\begin{bmatrix} U & \blacksquare & G \\ \downarrow & \cdot & \uparrow \\ G & \leftarrow & U \end{bmatrix}\begin{bmatrix} G & \blacksquare & A \\ \blacksquare & ⇙ & \blacksquare \\ A & \blacksquare & G \end{bmatrix}\begin{bmatrix} A & \rightarrow & U \\ \downarrow & ⇘ & \blacksquare \\ U & \leftarrow & A \end{bmatrix}\begin{bmatrix} G & \blacksquare & A \\ \blacksquare & \cdot & \blacksquare \\ A & \blacksquare & G \end{bmatrix}$$

$$\begin{bmatrix} U & \blacksquare & A \\ \uparrow & ⇙ & \blacksquare \\ A & \blacksquare & U \end{bmatrix}\begin{bmatrix} A & \rightarrow & U \\ \blacksquare & \cdot & \uparrow \\ U & \blacksquare & A \end{bmatrix}\begin{bmatrix} G & \leftarrow & U \\ \blacksquare & \cdot & \downarrow \\ U & \rightarrow & G \end{bmatrix}\begin{bmatrix} A & \blacksquare & U \\ \blacksquare & ⇘ & \uparrow \\ U & \leftarrow & A \end{bmatrix}$$

Annotation:

Only through our way of life and behavior do we create our living conditions, which either lead to an increase in the quality of life or to an increase in our physical, mental, and emotional ailments over time, which prevents us from fully regenerating ourselves, which we do deprive ourselves of the possibility of being able to keep our performance high for a lifetime and accordingly switch more and more into a passive and thus destructive way of life.

Naturally, vegetations, animals, and humans arise through the corresponding development of genetic material from the hydrogen isotopes: protium, deuterium, and tritium. That is also the reason why the DNA is transparent in the pictures.

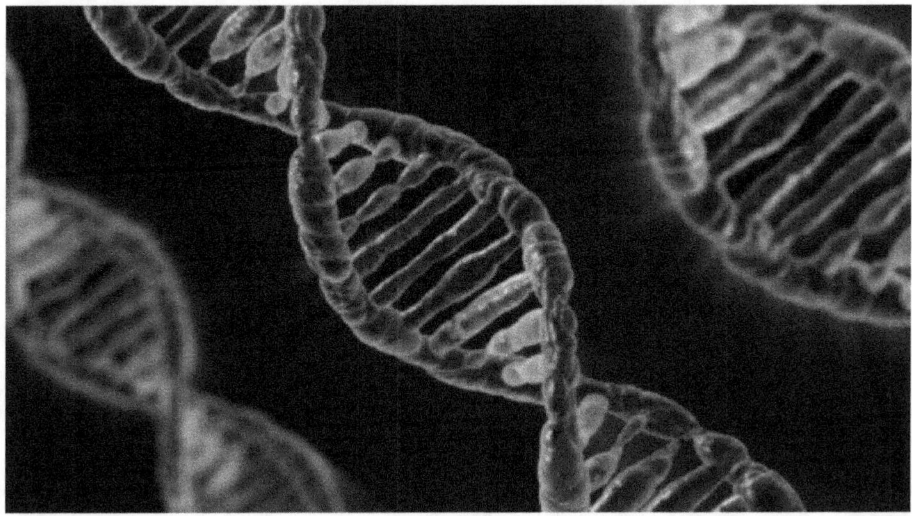

Based on this knowledge, one can also develop the amino acid bases as follows:

1) Guanine:

Guanine naturally consists of 100% tritium.

2) Adenine:

Adenine, therefore, consists of 100% deuterium.

3) Thymine:

Otherwise, thymine is 100% Protium.

4) Uracil:

However, uracil consists of 100% radium.

5) Cytosine:

Finally, it turns out that cytosine is composed of protium + radium.

Of course, you can also approach genetics in this way to develop corresponding mRNA active substances. But, since we want to be superior to nature, we must rely on daring theories and practices accordingly to put our stamp on the heart with our superiority.

I want to bring in the corresponding graphic representation right away.

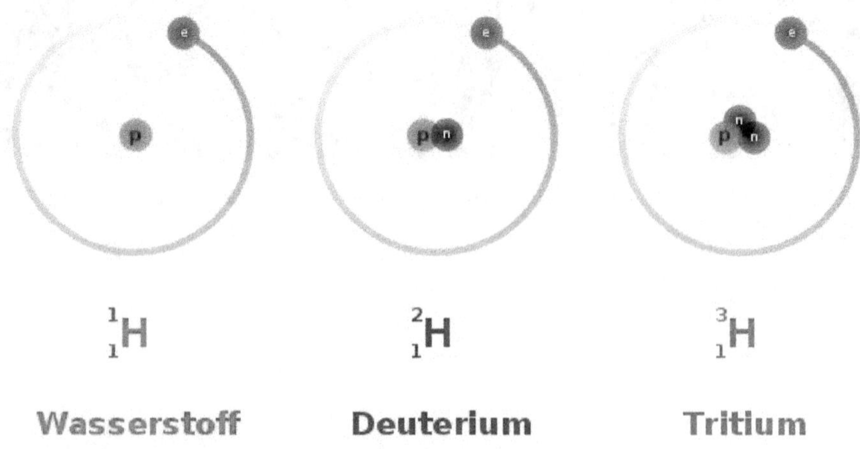

Quelle: Datei: Hydrogen Deuterium Tritium Nuclei Schmatic-de.svg - https://de.wikipedia.org

An error can have been seeing in the illustration above. Protium must have been using in place of the term hydrogen. In the book: "Formulas - Alles für Schule, Studium und Beruf" in Chapter 7 on pages 144 and 145, 2003 Tosa Verlag, Vienna, the following facts are stated: "In nature, hydrogen occurs as a mixture of isotopes (protium, deuterium, and tritium), the isotopes appear in different proportions."

The benefit of the hydrogen isotopes lies in developing more mental and physical performance and much older age. But of course, it allows hybridity to create from it.

Before I forget to mention, concerning the genome of coronaviruses (retroviruses, stem cells), the prickly protein consists of either thymine or adenine, from which the Y chromosomes then arise. The X chromosomes have therefore derived from guanine or cytosine. However, you can also produce Y and X chromosomes from any amino acid-base to obtain a species accordingly.

I can also add the following to the living conditions. First, clean air has obtained when the air has been cleaning, and the corresponding air purification device absorbs the CO2 in the room. Second, the CO2 is then converted into a liquid state in the air purifier to give the liquid CO2 corresponding hydrogen isotopes. Third, the CO2 can be released back into the environment after being cleaned by the isotopes. So, to speak, the CO2 is converted into hydrogen by the isotopes to be re-released into the environment in a gaseous state.

Incidentally, you can also heat with CO2, which means no need to import expensive gas.

At times I have thought of an RNA and DNA sequence that exhibits hybridity due to the nature of the hydrogen isotopes:

Law on the hybridity of genomes
A Tribute To Irma La Dulce

1) Version A:

$$\begin{bmatrix} T & \cdots & A \\ \vdots & {}^{1}_{1}H & \vdots \\ G & \cdots & G \end{bmatrix}\begin{bmatrix} A & \cdots & G \\ \vdots & {}^{2}_{1}H & \vdots \\ G & \cdots & A \end{bmatrix}\begin{bmatrix} G & \cdots & G \\ \vdots & {}^{3}_{1}H & \vdots \\ A & \cdots & T \end{bmatrix}$$

$$\begin{bmatrix} T & \cdots & G \\ \vdots & {}^{1}_{1}H & \vdots \\ G & \cdots & A \end{bmatrix}\begin{bmatrix} G & \cdots & A \\ \vdots & {}^{2}_{1}H & \vdots \\ A & \cdots & G \end{bmatrix}\begin{bmatrix} A & \cdots & G \\ \vdots & {}^{3}_{1}H & \vdots \\ G & \cdots & T \end{bmatrix}$$

2) Version B:

$$6'|T \quad A \quad AG \quad G \quad G|3'$$

$$3'|T \quad A \quad G|3'$$

$$3'|G \quad A \quad T|3'$$

$$3'|G \quad G \quad GA \quad A \quad T|6'$$

3) Version C, a derivative of version B:

6'|T A AG G G|3'

3'|T T TA G G|6'

6'|G A AT T T|3'

3'|G G GA A T|6'

4) Version D, a derivative of version A:

TA AG G G

TA ■G ■ ■

5) Version E, a derivative of Version D:

5'|T T AA A GG G G|9'

9'|G G GG A AA T T|5'

5'|T T AA A GG G G|9'

9'|G G GG A AA T T|5'

All versions represent mutations. The amino acid bases have the following task:

1) Reverse transcriptase:

The T cells from thymine are responsible for forming reverse transcriptase and forming red blood cells. The number of these changes must be recognized by how often the T appears in a row. The T cells are also responsible for the creation of Y chromosomes.

2) Adenosine dehydrogenase:

The B cells create the prerequisites for the ability to adapt to external circumstances. These are also responsible for the formation of adrenaline and supply the cells with oxygen. B cells are also required to enable capillary formation in the body. It also applies here that the number of A depends on the requirement.

3) Substrate phosphorylation:

The T-helper cells build up the immune system by producing testosterone. Therefore, the more T-helper cells there are, the more the G occurs, and accordingly, more muscle mass has built up.

To top it off, I took the liberty of making a copy of the study by Rossana Segreto and Yuri Deigin, whereby I only scanned the page with the corresponding sequences:

WILEY BioEssays

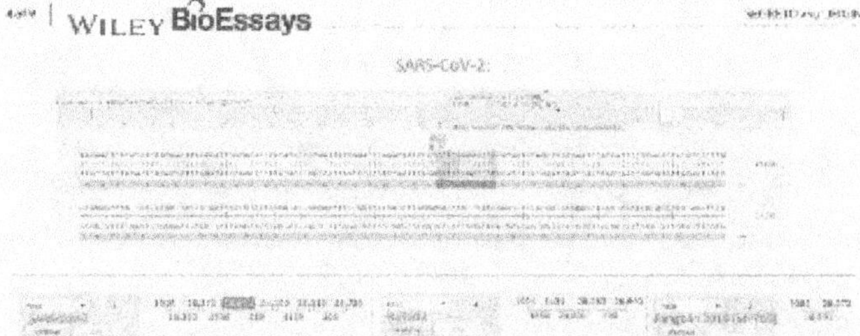

FIGURE 1 Nucleotide sequence of the S protein at the S1/S2 junction in SARS-CoV-2 (NC_045512.2) showing the furin cleavage site (in blue) that includes a FauI enzyme restriction site

FIGURE 2 Alignment of nucleotide and amino acid sequences of the S protein from bat-SL-CoVZC45 (MG772933.1) and RmYN02 at the S1/S2 junction site. No insertions of nucleotides possibly evolving in a furin cleavage site can be observed (in blue)

spike protein of SARS-CoV-2 of a cleavage site activated by a host cell enzyme furin, previously not identified in other beta-CoVs of lineage b[35] and similar to that of Middle East respiratory syndrome (MERS) coronavirus.[36] Host protease processing plays a pivotal role as a species and tissue barrier and engineering of the cleavage sites of CoV spike proteins modifies virus tropism and virulence.[37] The ubiquitous expression of furin in different organs and tissues have conferred to SARS-CoV-2 the ability to infect organs usually invulnerable to other CoVs, leading to systemic infection in the body.[38] Cell-cultured SARS-CoV-2 that was missing the above-mentioned cleavage site caused attenuated symptoms in infected hamsters,[39] and mutagenesis studies have confirmed that the polybasic furin site is essential for SARS-CoV-2's ability to infect human lung cells.[40]

The polybasic furin site in SARS-CoV-2 was created by a 12-nucleotide insert TCCTCGGCGGGC coding for a PRRA amino acid sequence at the S1/S2 junction (Figure 1). Interestingly, the two joint arginines are coded by two CGGCGG codons, which are rare for these viruses: only 3% of arginines are coded by CGG in SARS-CoV-2 or RaTG13, and CGGCGG in the new insert is the only doubled instance of this codon in SARS-CoV-2. The CGGCGG insert includes a FauI restriction site, of which there are six instances in SARS-CoV-2 and four instances in RaTG13 (and two in MP789). The serendipitous location of the FauI site could allow using restriction fragment length polymorphism (RFLP) techniques[41] for cloning[42] or screening for mutations,[43] as the new furin site is prone to deletions in vitro.[32,44]

A study by Zhou et al.[45] reported the discovery of a novel CoV strain RmYN02, which the authors claim to have natural PAA amino acid insertions at the S1/S2 cleavage site where SARS-CoV-2 has the PRRA insertion. However, upon close examination of the underlying nucleotide sequence of RmYN02 in comparison with its closest ancestors bat-SL-CoVZC45 and bat-SL-CoVZXC21, no insertions are apparent, just nucleotide mutations (Figure 2).

Therefore, SARS-CoV-2 remains unique among its beta-CoV relatives not only due to a polybasic furin site at the S1/S2 junction, but also due to the four amino acid insert PRRA that has created it. The insertion causes a split in the original codon for serine (TCA) in MP789 or RaTG13 to give part of a new codon for serine (TCT) and part of the amino acid alanine (GCA) in SARS-CoV-2 (Figure 3).

The insertion of the furin cleavage site in SARS-CoV-2 is not in-frame with the rest of the sequence, when compared with the MP789 and the RaTG13 sequences (Figure 3). Therefore, it is possible to exclude that such insertion could have originated by polymerase slippage or by releasing and rejoining, because insertion mutations generated by these mechanisms have been postulated to maintain the reading frame of the viral sequence.[46] The possibility that the furin cleavage site could have been acquired by recombination has been recently

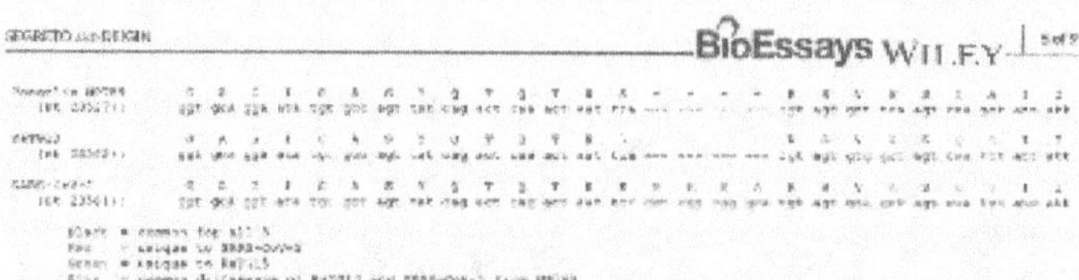

If I'm overdoing it with genetic pattern recognition, then I'd like to add another one. Even if it is so confusing, the whole genetic pattern should only show the permutation possibilities. Many virologists and geneticists always assume the same design, but this is not because humans have too many distinct differences. As mentioned earlier, if Americans had deciphered the human genome entirely, then with stem cell therapy alone, they could repair all kinds of mental and physical damage since mental illness only occurs because people lose their brain cells due to a lack of mental activity. As an act of charity, I will also send Ms. Merkel her genetic pattern, which will make her diamond even more effective.

Law of Mrs. Angelika Merkel

C TA GG AT C
T CG AA GC T
A GC TT CG A
G AT CG TA G
A GC TT CG A
T CG AA GC T
C TA GG AT C

While looking at some pictures in the Windows archive, I made an exciting discovery for myself. I will use the concept of vitamin B3 - read thiamine to show how one can read off a genetic

Vitamin B1

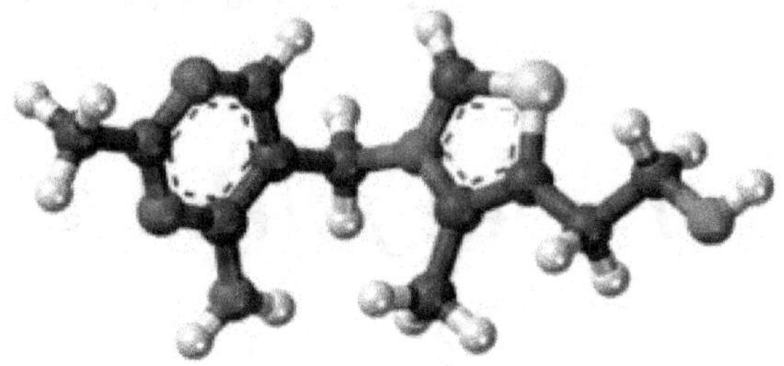

Thiamin

structure.

Based on the above picture, I try to read off a corresponding genetic structure. White stands for uracil, blue for adenine, red for thymine, yellow for cytosine, and black for guanine replaced these accordingly. Then, based on the thiamine or the structural model, one can see very well what peplomers are for it. The coronaviruses (retroviruses, stem cells) combine via the peplomers so that a related species emerges. It is also one of the reasons why I created the many genetic patterns because nobody knows how to transfer these identical genetic patterns to life, find out what effect or defect is behind them, and prove that genetics is still in its infancy. The thiamine contains the following structure, for example, with the two rings in my sights:

4'|*A G AG G G*|3'

3'|*G G GA G A*|4'

3'|*A G G G*|3'

3'|*G G G A*|3'

I've finally learned how to tell the difference between X and Y chromosomes in the RNA sequence. The 4' stands for the X chromosome and the 3' for the Y chromosome in the first ring. In the second example, there are only Y chromosomes marked with the number 3'. The mirroring of the RNA sequence, which I always use, generates a DNA sequence, checking whether the RNA sequence is only composed of X or Y chromosomes or both. To better recognize the difference at the molecular level, one should know - including me - that at least five elements have required creating an X chromosome and at least four aspects for a Y chromosome. Sometimes the molecular level should not be confused with the genetic level since the interpretation of genomes with the aid of structural models are subject to permutation and therefore permit many different genetic patterns.

If I've ever been there and read that BioNTech is planning to develop a malaria drug, you should watch out for sickle cell disease. I took all the articles about it from Wikipedia and only translated them into English.

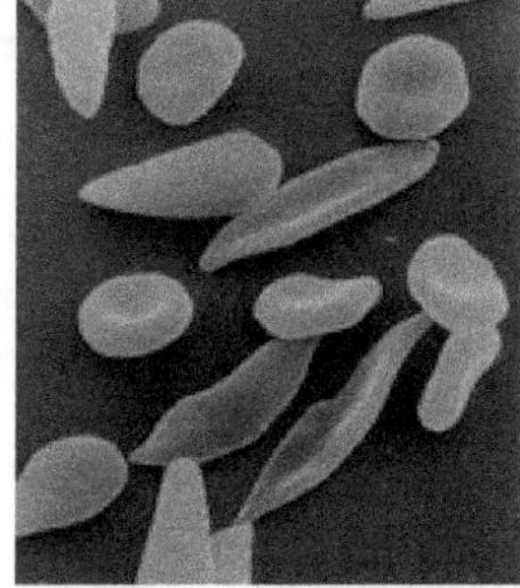

Quelle: Datei: Sicklecells.jpg - https://de.wikipedia.org

In the picture below, you can see the white areas very well. These white spots could also be coronaviruses.

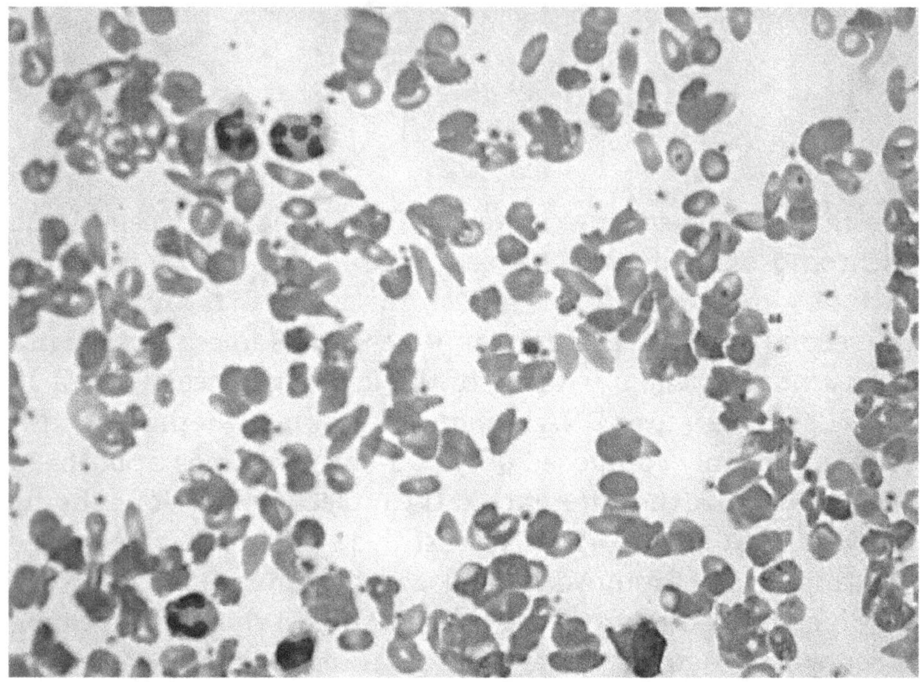

Quelle: Datei: Sikkel2.jpg - https://de.wikipedia.org

Those affected produce abnormal hemoglobin (sickle cell hemoglobin, HbS), which tends to form fibrils when there is a lack of oxygen. The red blood cells are deformed into sickle-shaped structures by the fibers they contain, clump together, and clog small blood vessels, causing inflammation. In the homozygous form, clumping and vascular blockage can lead to attack-like, painful, sometimes life-threatening circulatory disorders (sickle cell crises), leading to venous thrombosis, among other things. People affected by heterozygous diseases, in whom only one of the two hemoglobin genes has changed, are protected from the severe forms of malaria. As a result, the mutated hemoglobin gene is relatively common in malaria areas.

The destruction of red blood cells leads to severe chronic anemia (hemolytic anemia). Due to the tendency of hemoglobin S to polymerize and the sickle-shaped deformation of the erythrocytes, occlusions of small arteries occur with recurrent circulatory disorders. It leads to severe pain and damage to multiple organ systems: brain (ischemic stroke), splenic infarction, lungs (pneumonia, pulmonary hypertension), eye, heart and kidney failure, muscle, bone (osteonecrosis), or priapism. Life expectancy has reduced. Glomerulopathy with hyperfiltration occurs in up to a third of patients with a homozygous phenotype in childhood. Damage to the renal medulla leads to papillary necrosis, loss of the kidneys' ability to concentrate, and bloody urine (macrohematuria). Damage to the kidney corpuscles (glomeruli) leads to increased protein excretion in the urine (micro-and macroalbuminuria, nephrotic syndrome). In the histological examination, the predominant glomerular damage is focal segmental glomerulosclerosis. Proteinuria occurs in up to a third of patients' first decades of life, and terminal kidney failure in five percent.

Only homozygous carriers of the sickle cell gene show this severe form of the disease, in which all the hemoglobin is abnormal sickle cell hemoglobin (irregular hemoglobin). In heterozygous carriers, only about one percent of all erythrocytes have deformed. Symptoms are significantly worse when people are physically active or at high altitudes. The sickle shape of the erythrocyte forms when the partial pressure of oxygen is low. Under these conditions, the hemoglobin is fibrous (the solubility of hemoglobin in sickle cell anemia is 25 times less than the solubility of normal hemoglobin).

Symptoms can appear for the first time from around six months when the breakdown of fetal hemoglobin is already well advanced. They usually manifest themselves in what is known as a sickle cell crisis: External influences such as exertion reduce the oxygen partial pressure in the blood, and the sickle cells become hemolytic.

caused

Due to a point mutation in the HBB gene (c.20A> T) on chromosome 11, the amino acid glutamic acid has been replacing by valine at position six of the β-globin protein subunit of hemoglobin in sickle cell anemia. The designation of this variant in the official genetic nomenclature is HBB-p.E6V. The affected erythrocytes become sickle-shaped with decreasing oxygen partial pressure, easily get caught in the capillaries, and lyse quickly. Hemolysis releases hemoglobin, arginase, and free oxygen radicals. Free hemoglobin binds nitric oxide about 1000 times more strongly than intracellular, and arginase converts nitric oxide to nitrite and nitrate. Nitric oxide is an essential vasodilator, and the decrease in concentration leads to vasoconstriction and thus to circulatory disorders.

Sickle cell hemoglobin is referred to as HbS in contrast to HbA, the average hemoglobin of adults. In addition to HbS, heterozygous carriers of the trait also produce HbA in sufficient quantities to essentially maintain the function of the erythrocytes in these people.

Using sickle cell anemia as an example, the connection between a defect in a molecule and a disease was demonstrated for the first time in a famous work by Linus Pauling, Harvey Itano, and Seymour Jonathan Singer from 1949. The difference in the hemoglobin of both red blood cells has shown in gel electrophoresis, Itano performed. The authors already suspected differences in the amino acids, which Vernon Ingram confirmed in 1956, who also showed that the difference consisted in exchanging exactly one amino acid. The inheritance pattern of the disease was also clarified in 1949 by James Van Gundia Neel (1915–2000).

Inheritance

Sickle cell anemia is an autosomal codominant hereditary disease.

The genetic make-up of a healthy person contains the two incompletely dominant (A and S are codominant at the molecular level) alleles (AA) for hemoglobin A. As a result, its red blood cells are always elastic.

A carrier (carrier) with the genotype AS (= heterozygous) contains both the allele A and the mutated allele S, which causes the changed hemoglobin S. Its red blood cells to have HbA and HbS in a ratio of 1: 1. Under normal conditions, the red blood cells show no changes, and the disease does not break out. Only when there is an extreme lack of oxygen do the red blood cells deform into sickle-shaped structures, impairs the blood flow to the organs.

A carrier of the genotype SS (homozygous) only produces the modified HbS. Even under a physiological lack of oxygen, as it is, for it. B. in the capillaries of oxygen-consuming organs, there is a substantial deformation of the red blood cells. They lose their elasticity and easily get caught in one another. It leads to a closure of the capillaries. Under normal conditions, the hemoglobin in the red blood cells is always finely divided. With decreasing pH and oxygen levels in the blood, hemoglobin molecules clump together to form rod-shaped, crystalline structures in HbS. As a result, the erythrocyte is deformed into a sickle shape and loses its elasticity.

diagnosis

The diagnosis of sickle cell anemia has initially based on anamnestic, whereby the origin and other cases of the disease in the family have been inquiring about it. Then, clinically based on the symptoms and finally in the laboratory, where hemolytic anemia can appear in a blood count, and the typically shaped drepanocytes appear when the blood is examined under the microscope - especially if the blood has been storing under the exclusion of oxygen for 24 hours (sickle cell test). In addition, electrophoresis of the hemoglobin can conclusively identify the changed molecules. Finally, the gene segment for hemoglobin can also be examined in a restriction analysis, and the point mutation can have been shown at the DNA level.

Sickle cell test

The blood to have been examining is mixed with EDTA and stored for 24 hours under the exclusion of oxygen. As a result of the lack of oxygen in the erythrocytes, the sickle shapes of the cells arise, which can have been quickly recognizing under the microscope at 40x magnification. In addition, the sickle cell formation effect can have been accelerating by adding sodium disulfite (sodium metabisulfite).

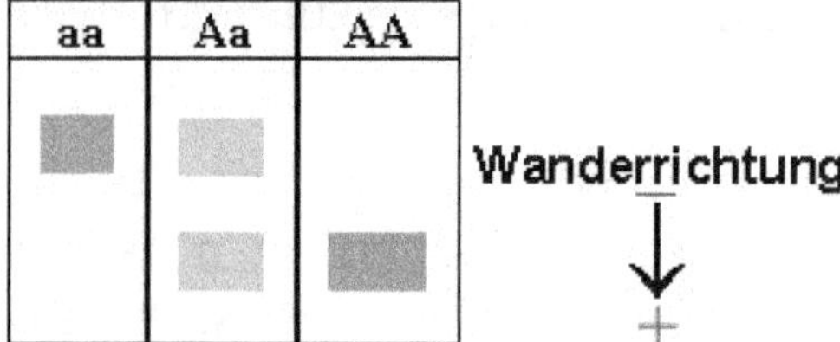

Quelle: Datei: Elphor.png - https://de.wikipedia.org

Significance of sickle cell anemia in malaria

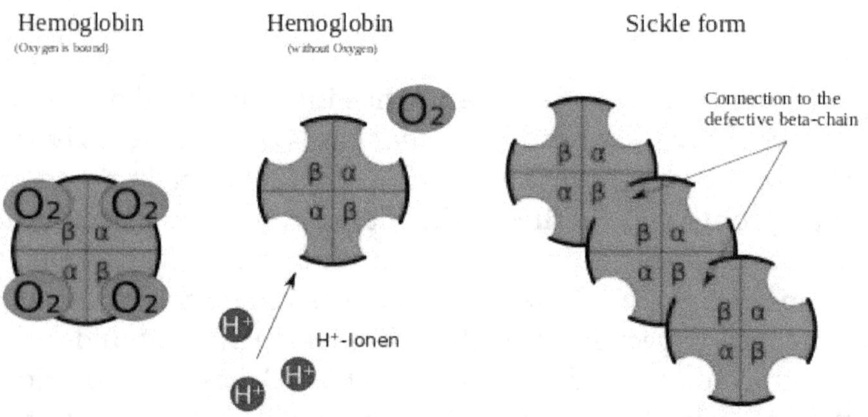

Quelle: Datei: Hämoglobin bei der Sichelzellenanämie. Svg - https://de.wikipedia.org

There may be a selective advantage of heterozygous carriers in malaria infections. The malaria pathogen is transported to or in the erythrocytes during part of its development cycle. The hemoglobin of people with the heterozygous form of HbS leads to a sickle-like deformation of the red blood cells by reducing the oxygen saturation of the hemoglobin under extreme conditions, which are then broken down or clumped in the spleen and then perish. One hypothesis says that those cells infected by plasmodia would deform themselves through the influence of the merozoites or trophozoites, even without reducing pressure, and would be recognized and broken down by the spleen.

Another hypothesis is the direct killing of the parasites because the sickle cells produce more oxygen radicals. As a result, superoxide anions and hydrogen peroxide have formed, and both compounds are toxic to the parasites.

Another theory says that if the plasmodia that cause malaria attack erythrocytes, the microbes release acids as waste products of their metabolism over time. The hemoglobin now releases the oxygen, stimulated by H + ions (shift to the right of the oxygen-binding curve). The sickle shape mainly affects the deoxy body of the erythrocytes. So, the infected cells quickly become sickle cells, broken down in the spleen and the microbes. It explains the resistance of the carriers of sickle cell anemia to malaria (see figure).

The last theory says that hemoglobin polymers are formed in the process, leading to the direct killing of the parasites.

therapy

Approaches to amplifying HbF gene expression in adolescents and adults are currently have been investigating.

Hydroxyurea can induce the formation of HbF. Red blood cells with a high proportion of HbF do not form sickle cells and are less likely to break down and cause small vessels to block. Thus, treatment with hydroxyurea can reduce the frequency of vascular occlusions, alleviate chronic organ damage, and extend survival. It has recently been showing to children in sub-Saharan

Africa.

Furthermore, adoptive cell transfer has been investigating.

On October 12, 2016, a treatment option based on a change in the affected genes using the CRISPR / Cas's method was published. With the help of gene scissors, the researchers replaced the pathogenic mutation with the correct DNA bases. For the first time, enough healthy blood cells have been generating to cure patients with this method in the future, as reported by researchers working with Jacob Corn from the University of California, Berkeley. It is still too early to speak of a workable solution, but the first step is to combat the causes instead of treating the symptoms.

Other drugs are the oxygen affinity modulator voxelotor and the monoclonal antibody crizanlizumab.

Corresponding hemoglobin anemia can also occur in HIV, Ebola, leukemia, etc. So perhaps sickle cell disease also has something to do with SARS. Sickle cell disease can also be an indication that there is too much nitrogen in the blood. Sickle cell disease has been triggering because coronaviruses introduce nitrogen or toxins into the cell nucleus via the prickly protein by attaching themselves to the cell with the help of the peplomers, as they also connect via the peplomers. Incidentally, the coronaviruses need their peplomers to be able to produce a polymerase through them.

Now, once and for all, what I've come up with so far is ready to be said. The body's coronaviruses (retroviruses, stem cells) are never detectable in one's blood. If coronaviruses are still noticeable in the blood, they come from the outside through fungi or other animals. Alien coronaviruses, i.e., coronaviruses that do not belong to their species, always produce pox-like symptoms, leading to inflammation within an organ. I am also right in saying that there is a difference between peplomers and telomeres. The telomeres have located inside the peplomers. The telomeres have been used for defense and to puncture the cell membrane to absorb the contents of a cell via the peplomers. The principle is also crucial, e.g., in the case of mosquitoes, which first penetrate the skin with the help of the telomeres, then the telomere is retracted so that the blood has withdrawn via the peplomer. The coronaviruses, so to speak, ensure that living beings display specific defensive and eating behavior.

REVERSE TRANSCRIPTASE LAW

A Tribute to Patric U.B. Vogel

The genetic code consists of a sequence of nucleic acid sequences. Each genome has made up of a chain of nucleic acid sequences called DNA. A single chain of a nucleic acid sequence is called RNA. To generate a DNA from a single chain of a nucleic acid sequence, this single chain of a nucleic acid sequence is converted into a protein and passed on as a ribosome to the respective single chain, i.e., single-strand, of a nucleic acid sequence, i.e., appended so that from a single chain, i.e., Single strand, a nucleic acid is converted into two intertwined chains of a nucleic acid sequence, i.e., double-strand. This process is called reverse transcriptase. The following should also have been mentioning; If a single chain of a nucleic acid sequence has converted into a protein, one speaks of an mRNA.

Example:

a) Single strand (RNA or also called sRNA):

G AT

b) <u>mRNA (ribosome or protein-coding enzyme):</u>

$$G \quad AT \quad C$$

is converted into a protein.

c) <u>DNA (also known as ssRNA):</u>

With the help of reverse transcriptase, we get:

$$\begin{bmatrix} G & \rightarrow & A \\ \uparrow & \ddots & \downarrow \\ C & \leftarrow & T \end{bmatrix} \begin{bmatrix} T & \rightarrow & C \\ \uparrow & \ddots & \downarrow \\ A & \leftarrow & G \end{bmatrix}$$

d) <u>Note to c):</u>

If I were to only use guanine as mRNA, for example, then the following would result:

$$\begin{bmatrix} G & \rightarrow & A \\ \blacksquare & \nearrow\!\!\nearrow & \downarrow \\ G & \blacksquare & G \end{bmatrix} \begin{bmatrix} T & \rightarrow & C \\ \uparrow & \nearrow\!\!\nearrow & \downarrow \\ G & \blacksquare & G \end{bmatrix}$$

There are, of course, many other ways in which, for example, a coronavirus can have been manipulating, or other genetic manipulations can have been bringing about if, for example, the M protein of the coronavirus has been defining as RNA, the spike protein as mRNA and the E protein as DNA.

MECHANOTRANS-DUCTION IN VIROLOGY

A Tribute to Mai Thi Nguyen-Kim Leiendecker

And Angelika Merkel

Suppose you manage to activate an adenosine dehydrogenase in the body using adenoviruses as soon as coronaviruses penetrate the cells. In that case, you trigger a mutation in the coronaviruses that prevents them from leaving the cells. It has achieved when adenoviruses are cultivated in deuterium to administer them subsequently. The deuterium in the adenoviruses ensures that the so-called sickle cell disease develops in the coronaviruses, which causes the spike proteins - i.e., peplomers - to die off. In addition, deuterium prevents the formation of the autoimmune disease caused by the coronavirus spike proteins. It should also have been noting that the first primates arose from the adenoviruses and that today's chimpanzees are the descendants.

VECTOR VACCINES - GENERATION OF ADENOSINE DEHYDROGENASE AND ATP SYNTHASE BY LYSIS

A Tribute to Dr. Anthony Fauci

1) An egg cell is supposed to serve as the virus envelope. The thymine already provides this to us since it consists of thymine.

2) Any type of virus can now be introduced into the egg cell. If this egg cell were to have been fertilizing, there would be an immunization of the resulting generation and a genetic change.

3) The unfertilized egg cell with the corresponding viruses inside can be inoculated.

4) Whether the viruses had weakened or inactivated before they were implemented in the egg should depend on the goal achieved.

5) If one wants to obtain mechanotransduction, the property of which, within virology, is to cause codon optimization or codon de-optimization.

6) Codon optimization is used for selection and can only be achieved if active viruses are implemented in the egg to eliminate genetic dispositions.

7) Codon de-optimization, on the other hand, is used for assimilation, which has achieved when inactive viruses are implemented in the egg cell also to remedy genetic dispositions.

8) In both cases, genetic dispositions are eliminated because codon optimization produces an adenosine dehydrogenase that leads to the formation of antigens so that the immune system remains immunogenic throughout its life.

9) But codon de-optimization only creates preliminary immunogenicity, which means that regular vaccination is indispensable.

10) For this reason, codon de-optimization is responsible for the so-called placebo effect, i.e., ATP synthase due to the lysis.

11) Codon de-optimization is a prerequisite for the transience of one's own life - in other words: the aging process.

12) However, there is a difference in whether codon de-optimization has been causing by viruses, i.e., living conditions or personal misconduct.

13) Alcohol - no matter in what form and quantity - destroys the DNA. The evidence is that alcohol has a negative impact that destroys the body, mind, and spirit.

14) The proof that I am right that inactivate vaccines only produce a placebo effect, i.e., ATP synthase through the lysis that has triggered by spike protein, is in the book: “COVID-19: Search after a vaccine” by Patric UB Vogel, p. 28/29, only available in German, Springer Spectrum, supplied: “After intensive efforts, two different types were approving in the USA in 1963, a live vaccine and a whole virus inactivated vaccine. Both have been

using to prevent measles. Unfortunately, however, the inactivated vaccine has withdrawn after a few years (Hendriks and Blume 2013). The reason was that despite the triple vaccination, this vaccine only stimulated a relatively short-term antibody response and even caused severe and atypical, i.e., asymptomatic disease courses in some of the people vaccinated with it when they were infected (Nossal 2000)."

15) The mRNA vaccines are also known as split vaccines. The mRNA vaccines consist of membrane fragments with embedded proteins because the m stands not only for "messenger" but also for membrane, and the viral proteins each form an RNA.

16) The mRNA is crucial for developing the corresponding ribosomes needed to produce live vaccines.

17) ATP synthase through lysis manifests itself in the formation of lysosomes, which are needed to produce a lipid membrane envelope.

18) Adenosine dehydrogenase, on the other hand, manifests itself in the formation of ribosomes, which have required to be able to produce spike proteins.

19) To find out whether a vaccine is working or causing side effects, I recommend first giving a whole virus inactivated vaccine, then a vector or mRNA vaccine, or the like.

20) If, for example, you take a whole virus active vaccine after an actual virus inactivated vaccine, the active virus can no longer dock onto the host cell, let alone penetrate the cell, since this virus already occupies the host cell. Should it turn out differently, the deuterium ensures that the active virus can no longer multiply.

21) the Whole virus inactivates vaccination has also been recommending for those affected who have already broken out. Viruses must have been cultivating in deuterium for an active vaccine. I also recommend not increasing the dose if immune senescence occurs. It is much more pleasant and safer to in-

crease the frequency of vaccination.

22) Coronaviruses (retroviruses, stem cells) can only multiply and grow into an organism if they combine. As soon as a coronavirus is inside a cell, it is no longer possible to reproduce, as these go into a state of dormancy to mutate. Otherwise, one can test all kinds of vaccines for effectiveness and side effects with the help of chicken eggs. If vaccines are effective, the chick will not harm the chick if infected. However, if side effects occur, the chick will have corresponding genetic dispositions, expressed by related defects, external or internal. You can also find out to what extent external and internal genetic changes occur in the chick and chicken if you fear that this might happen.

QUESTIONS TO ASK:

1) Do coronaviruses multiply by themselves, i.e., without combining with other coronaviruses?

2) Or do the coronaviruses need other coronaviruses to combine with them so that reproduction and growth are possible?

3) Does the deuterium prevent multiplication, or does the deuterium prevent the coronaviruses from connecting?

4) The question also arises: If a coronavirus has already occupied a host cell, can it still be occupied by other coronaviruses?

5) How does a coronavirus penetrate a host cell?

6) Which mechanism allows an intrusion?

7) Has penetration into a cell been observed directly under laboratory conditions?

8) To what extent do computer animations and graphic representations reflect reality if the original recordings cannot be used?

9) Coronaviruses - no matter what type - continuously adapt to the host cell and not the other way around. It is also why coronaviruses - no matter what kind - do not influence genetics because the host cell would have to adapt to the coronavirus.

10) The spike proteins are nothing more than phages that attach themselves to a spore and inject something into the spore, causing what is known as phagocytosis.

11) To find out whether coronaviruses have a genetic influence, you can inject chicken eggs, i.e., each egg, with a different coronavirus, and then let them hatch and grow up. Then, you will

see whether a genetic change can have been detecting or not.

12) One can also learn to understand the principle of phages by introducing only amino acid bases into chicken eggs, then hatching them and letting them grow up.

13) Despite all the ideas I have developed and the possibilities, you can see the scientific consensus regarding the corona crisis. The danger is expected from the spike proteins that many underestimate - I also have the threat underrated. Spike proteins release a protein to a host cell, leading to pimples or warts, for example, forming on the skin's surface. In this regard, spike proteins trigger a septic behavior so that organs have impaired functioning, whereby the protein, e.g., obtained from pus pimples, is highly inflammatory. For this reason, I am lucky that I have watched the relevant videos again and again until after a certain period, it dawns on me. Accordingly, I can associate suppuration, i.e., sepsis, with coronaviruses and their spike proteins.

14) With the help of ATP synthase through lysis, i.e., the biological process of electrolysis, whole virus inactivated vaccines can reactivate the administered viruses.

15) As mRNA vaccines, one can use a whole virus active vaccine, e.g., adenoviruses or SARS-CoV-1 viruses or at least another coronavirus, which only docks on a cell without penetrating it, which makes the supposedly "more dangerous." "Virus docks on to the" weaker "virus, creating an ATP synthase through lysis, which forces the more fragile virus to decouple from the cell.

16) A protein consists only of the virus envelope and the cytoplasm. Thus, for example, an mRNA vaccine can be produced using appropriate ribosomes and a vector vaccine using lysosomes.

17) I can recommend the following YouTube video, which is

only available in German or French: "Corona: The race for the vaccine," documentary, Arte, ARTE de.

19) The biomass - i.e., protein, i.e., DNA - from coronaviruses has only obtained the combined coronaviruses. Thus, a coronavirus mutation occurs, for example, when different coronaviruses combine. And the biomass is also accepted by knowing the required amino acid bases.

20) As soon as a virus penetrates the cell and multiplies, phagocytosis starts to form. Phagocytosis of coronaviruses is required for selection and assimilation since coronaviruses can absorb something via the peplomers and release something into the environment.

21) Phagocytosis corresponds to the piezoelectric effect.

22) The genetics of the SARS-CoV-2 virus are taken as mRNA, which serves as a culture medium, and adenoviruses have been cultivating on this culture medium. Then you have a whole virus active vaccine, which ensures that the SARS-CoV-2 viruses can no longer occupy cells because the vaccine with adenoviruses already inhabits these. Should there be a connection, a new genome will develop without the disadvantages for vaccinated persons. The adenoviruses, then the coronavirus, select the coronavirus to convert its spike proteins into cytoplasm assimilate its virus envelope even to spike to produce proteins.

23) If the virus envelope consists of adenine, then the guanine becomes the peplomer. But if the virus envelope consists of thymine, then the guanine creates the telomeres.

24) To be able to isolate individual components of viruses, electrophoresis is required.

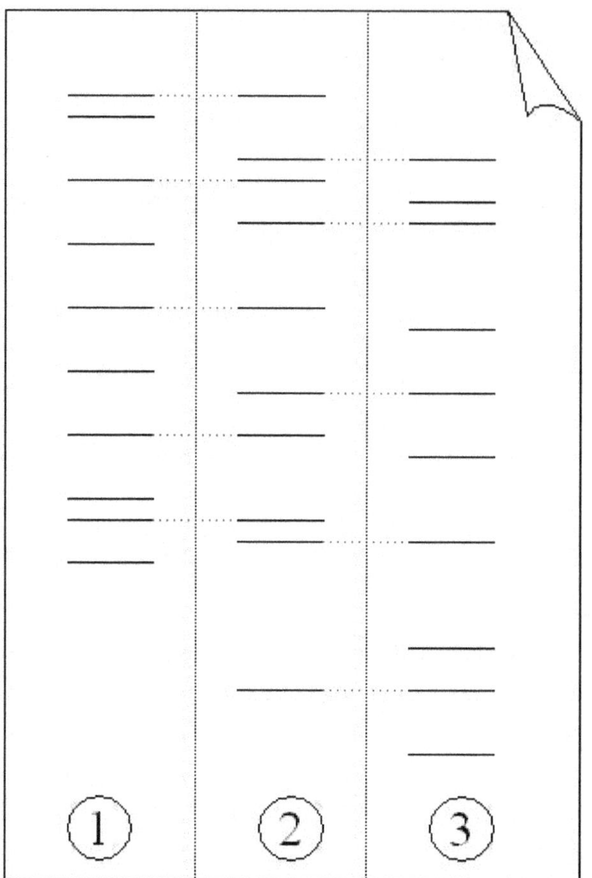

Quelle: Datei: PCR fingerprint.png - https://de.wikipedia.org

"Elektrophorese mit DNA-Fragmenten, die über PCR gewonnen wurden;

(1) ist der Vater,

(2) das Kind,

(3) die Mutter."

Quelle: Elektrophorese - https://de.wikipedia.org

25) Electrophoresis (outdated cataphoresis) describes the migration of charged colloidal particles or dissolved molecules through an electric field. The pioneer of electrophoresis was Arne Tiselius (1937). The breakthrough came after Oliver

Smithies found in 1955 that starch gels were very suitable for electrophoresis (later replaced mainly by acrylamide, for example).

HUMAN CORONAVIRUS (MODIFIED)

1) Virus envelope made from thymine.

2) Guanine spike protein.

3) M protein from adenine.

4) E protein from cytosine.

5) Virus RNA + N protein from uracil.

6) Do we get an antigen from points 1) to 5)?

METHOD OF INTERPRETING GENETIC STRUCTURES THAT AIMS AT THE PRINCIPLE OF ELECTROPHORESIS

In Dedication to Arne Tiselius

And Oliver Smithies

To better understand the way of interpretation, I will use two different mRNA sequences.

The first mRNA sequence is: 3'|GGGTG|5'

The second mRNA sequence is: 3'|TTTGT|5'

The following statements can have been making about both mRNA sequences: 1) The first mRNA sequence shows us that the (virus) envelope consists of guanine and the cytoplasm of thymine. 2) The second mRNA sequence, on the other hand, shows us that the (virus) envelope consists of thymine and the cytoplasm of guanine. The reason for this hypothesis (speculation, assertion, etc.) is that four identical components of a virus form the virus envelope, three similar features of a virus include the Y chromosome, i.e., the spike proteins and two identical components of a virus initiates the cytoplasm.

It gives us the following result:

1) For the first mRNA sequence

$$\begin{bmatrix} G & \cdots & G \\ \vdots & T & \vdots \\ G & \cdots & G \end{bmatrix}$$

2) For the second mRNA sequence

$$\begin{bmatrix} T & \cdots & T \\ \vdots & G & \vdots \\ T & \cdots & T \end{bmatrix}$$

The son supplies the first mRNA sequence, and the daughter provides the second mRNA sequence. From this, one can infer who is related to whom.

3) Now, we determine the DNA sequence of the father and mother.

a) first that of the father:

$$\begin{bmatrix} G & T & G \\ T & G & T \\ G & T & G \end{bmatrix}$$

In the case of the father's DNA sequence, the guanine represents the spike protein. But since the spike protein is nothing more than phages which, in the case of the father, inject guanine into the interior of a spore. Thus, in the case of the father, the thymine represents the virus envelope.

b) next to the mother:

$$\begin{bmatrix} T & G & T \\ G & T & G \\ T & G & T \end{bmatrix}$$

In the mother's DNA sequence, the spike proteins consist of thymine, and the virus envelope consists of guanine. The spike proteins, i.e., phages, attach themselves to the spore and inject thymine into the interior of the spore.

4) And together, we get the following genome:

$$\begin{bmatrix} G & T & G \\ T & G & T \\ G & T & G \end{bmatrix}\begin{bmatrix} T & G & T \\ G & T & G \\ T & G & T \end{bmatrix}$$

5) In addition, it should have been mentioning that the injection

of guanine by the father creates telomeres and the infusion of the mother peplomers.

Conclusion:

At this point, I would end this topic. I chose the title above because there are moments when the tag is appropriate but also not. We humans are and will remain the most significant cause of illness precisely because we tend to misbehave. If we crap about the environment because we like to shit about each other, we will still reap many diseases because nature strikes back and then shit all over us. We treat animals like dirt and wonder when we eat dirt. We consciously poison our plants and wonder if we have been poisoning. Our diseases become less the more we give our way of life and behavior a higher quality. But if we should not do it, then even vaccinations will not help us anymore. The topics of virology, biology, chemistry, etc., and the developments so far have placed a lot of stress on me alongside my job.

Of course, I am not usually the one that handles letting others do their job, but at least I could get an impression of what it is like to be a student. To what extent I have correctly interpreted the issues relating to Corona or not, others must decide. Contradictions must always have been accepting since we humans will always find ourselves in disclaimer depending on the existing scriptures. But without contradictions, there is no progress, and it will always be and always will be. Am I entitled to express criticism? Yes, I am because I can do so if experts' statements contradict each other as taxpayers and responsible citizens. It creates the impression that no one has a clue, but everyone has an opinion. I developed my ideas from the background to show how I would approach something. Whether all of these come true or not is always a question of development, which also applies to the technical side. The work has been translated using Google and reviewed by Grammarly. Should misunderstandings arise anyway, give a shit about it. I wish everyone involved in my work a speedy recovery in their work. Over the months, I have developed different points of view so that nothing must be left to chance. If you have no experience like me, you should consider all the possibilities that allow everyone to develop their concept. I was also able to experience how difficult it is to create a unified vision from which ideas can arise. In case you were wondering why I included a lot of people I do not know personally? Quite simply because I am just a fan.

I almost forgot the following. By dealing with virology due to the Corona crisis, I discovered a lot of inconsistencies. For me, all scientists have now degenerated into notorious liars because they do not work professionally or not at all but only work with their mouths. Photographs taken under an electron microscope are still missing from most viruses and bacteria. There is not even a clear definition of the symptoms that occur for most diseases, and if so, the various conditions cannot have been distinguishing from one another. Too often, flu symptoms have advanced without inflammation of the mucous membranes being present or proven. All the quick tests are screwed. The whole pandemic has built on smoke and mirrors, but nothing more. And all those who have reported to the public on this topic so far have only pursued a disinformation policy to bring their conspiracy theories into the world. Virologists are to blame if they only utter superficial banter. HIV, Ebola, and many other diseases do not exist because, as already mentioned, there are no electron microscopic images, and the symptoms have not been determined and proven. The poorer countries only serve as test areas for vaccines and drugs. Since the people there have very little or no education, we can manipulate them more easily to fool them into wanting to help them, which we don't want. If we wanted to help people, we would improve their living conditions and not sell some placebo shit as a supposed miracle cure.

www.ingramcontent.com/pod-product-compliance
Lightning Source LLC
Chambersburg PA
CBHW070642220526
45466CB00001B/266
9798546847715